D0916894

The Soviet Energy Balance

Iain F. Elliot

The Praeger Special Studies program—utilizing the most modern and efficient book production techniques and a selective worldwide distribution network—makes available to the academic, government, and business communities significant, timely research in U.S. and international economic, social, and political development.

The Soviet Energy Balance
Natural Gas, Other Fossil Fuels, and Alternative Power Sources

Praeger Publishers   New York   Washington   London

PRAEGER SPECIAL STUDIES IN INTERNATIONAL ECONOMICS AND DEVELOPMENT

Library of Congress Cataloging in Publication Data

Elliot, Iain F
    The Soviet energy balance.

    (Praeger special studies in international economics
and development)
    1. Power resources—Russia.   I. Title.
HD9502.R82E45   333.7   73-187398
ISBN 0-275-08930-4

PRAEGER PUBLISHERS
111 Fourth Avenue, New York, N.Y. 10003, U.S.A.
5, Cromwell Place, London SW7 2JL, England

Published in the United States of America in 1974
by Praeger Publishers, Inc.

Printed in the United States of America

CONTENTS

LIST OF TABLES, FIGURES, AND MAPS

The Soviet Energy Balance

The importance of energy resources in determining a country's economic and political strength is now generally recognized. The United States and the Soviet Union are the world's main producers and consumers of energy, and although up-to-date information on the United States is readily available in English, this is not the case with the USSR. It is not sufficient merely to translate Soviet publications on fuel and power. Even with such factual subject matter, it is usually necessary to interpret material, making allowances for political pressures which can lead to exaggeration and misrepresentation. It is hoped that this study, based on the latest Russian sources, will help to satisfy the need for current data on what is not only a vital sector of the Soviet economy, but also a decisive factor in international relations.

Worldwide demands for energy are expected to continue rising to such an extent that in the last three decades of this century more energy may be required than the total amount consumed before 1970. In North America, traditional sources of power have already been recognized as inadequate, and alternative sources are now being developed. Oil and liquefied natural gas imports from the Middle East are subject to the restrictions imposed by the Arab nations, in order to deter U.S. support for Israel should conflict be renewed. Imports from the USSR, however, are unlikely to provide an easy alternative. There has been strong opposition in the United States to expanding trade with the Soviet government until it puts into practice such international conventions as the right of every citizen to travel freely abroad.

The Arab-Israeli confrontation has exacerbated political disarray in oil-hungry Western Europe, where several countries have been negotiating individual trade agreements to obtain oil and gas from the USSR. On the other hand, Eastern European nations, which are economically and politically dependent on the Soviet Union, are finding it increasingly necessary to supplement energy supplies from the USSR by importing oil from the Middle East.

Japan, the world's third industrial power, has to import almost all its fuel requirements, and is no longer prepared to rely completely on the Middle East for its oil. Although Soviet-Japanese cooperation could prove of great benefit to both countries, Japan is concerned about the consequences for its relations with China should Japanese capital and technology be used to develop the potential of Siberia and to im-

3

prove transport systems near the sensitive Sino-Soviet border.

The USSR has been expanding its influence significantly in the developing countries. Soviet experts have helped to exploit gas fields in Afghanistan and oil fields in Iraq. Considerable political importance is attached to Soviet oil exports to India, and Cuba is almost totally dependent on the USSR for its oil supplies.

In all the above examples, crucial questions arise concerning Soviet energy exporting potential, which can only be solved when recent information on reserves, production, and consumption are available. Such data are clearly also vital for any discussion of future Soviet industrial growth and regional development. The object of this study is to contribute to the debate on those and other related problems by providing some of the necessary factual material.

The first chapter traces the changes in the Soviet energy balance in the twentieth century, showing how, later than in many other industrial states, coal gave way to oil as the major source of energy and how natural gas and nuclear power are expected to provide a steadily rising share toward the year 2000. The rapid growth in the production of coal, oil, gas, peat, and oil shale which has marked recent decades is discussed, and the rates at which the output of each fuel has been increased are compared.

Separate chapters are devoted to natural gas, oil, coal, peat, and oil shale. Location of reserves, main extraction areas, production statistics, and transport methods are described for each fuel.

The seventh chapter briefly describes various sources of power other than fossil fuels, and in discussing electricity generation, introduces the complex subject of energy utilization. This is treated in greater detail in Chapter 8, which also deals with regional variations in fuel production and consumption. Trends in Soviet energy exports are described.

Several excellent works on Soviet energy resources have already been published in English. Unfortunately, the energy situation changes so rapidly that books published only a few years before can be quite misleading. Moreover, to the author's knowledge, no other work in English has attempted to discuss all the major sources of energy now being utilized in the USSR. As the centralized Soviet economy permits (at least in theory) the planning of a single integrated energy system for the whole country, no one resource should be considered in isolation. It does not make much sense to analyze Soviet oil export potential, for example, without also carefully examining the possibility of using coal, natural gas, local fuels, or other sources of energy to replace oil for certain requirements within the USSR.

This study is based almost completely on Soviet sources, and all data are presented in the metric system in accordance with the original Russian texts. To keep Soviet progress in energy production in perspective, comparisons have been made with the situation elsewhere, and in the United States in particular. United Nations publicacations have been the principal sources of comparative statistics.

The main sources have been listed in the bibliography and are given in detail in the notes which follow each chapter. They may be divided into six main groups: (1) general background material—encyclopedias, secondary works on history, geography and economics; (2) specialist works on energy resources and their utilization; (3) technical journals for the fuel and energy industries; (4) statistical yearbooks; (5) newspapers and periodicals; and (6) weekly economic reports on the USSR published by the monitoring service of the British Broadcasting Corporation. The last-named source has been particularly useful for updating regional production statistics.

Since 1955 there has been no shortage of detailed and accurate Soviet data on energy production and reserves, the main exception being material on oil resources, which is never published as it is still on the secret list. Since most of the sources consulted were not written for propaganda purposes but for the information of other Soviet experts, there has seemed little reason to doubt the validity of the compiled data, and in fact few inconsistencies have been found. Possibly the main difficulty has been one of terminology; for instance, the expression "tekhnologicheskoe toplivo" may simply be translated as "technological fuel"—but this has very little meaning for most Western experts and some further explanation must be attempted, even at the risk of slightly changing the original sense. In the tables, literal translations of the Russian headings have been preferred, although occasionally an explanatory note has been added.

The author wishes to express his appreciation to the librarians of Bradford University, Brighton Polytechnic, and the National Lending Library for Science and Technology. The constructive criticism and constant encouragement of Dr. V. S. Balashov of Bradford University have been indispensable for the completion of this work. Chapter 7 in particular owes much to his useful advice. Mr. W. J. Webb of Brighton Polytechnic read the manuscript and made many helpful suggestions. Mr. D. M. Elliot of Dumbarton Academy rendered invaluable assistance in the preparation of the maps. The author's sincere thanks go to all the above, and to his wife and parents for their patience and support in the years of research and writing.

The publishers have prepared the manuscript rapidly yet painstakingly; needless to say, any mistakes which remain are the sole responsibility of the author.

THE SOVIET ENERGY BALANCE

The Soviet Union has less reason than most industrial powers to fear overpopulation and the exhaustion of the natural resources within the limits of its own political boundaries. In 1973, with a population of 250 million, less than 7 percent of the total population of the earth, the USSR occupied a territory of 22.4 million square kilometers, about one-sixth of the inhabited area of the earth. It could claim 57 percent of the world's resources of coal, 45 percent of its natural gas, 60 percent of its peat, 46 percent of its oil shale, 12 percent of its potential hydroelectric power, and 37 percent of its oil-bearing area.[1]

Although these figures may be disputed, it is understandable that until recently the policy of the USSR in constantly increasing its consumption of energy was seldom questioned. Soviet planners concentrate on the methods by which further output can best be achieved, and graphs showing a steady rise in the amount of energy produced in the USSR are presented regularly to the Soviet citizen as one of the major indicators of technological progress. Every schoolchild in the USSR knows Lenin's maxim that "Communism is Soviet power plus the electrification of the whole country." When the press publishes statistics on the fulfillment of industrial norms, the list is invariably headed by the ministry for power production, followed by the fuel ministries.

This focusing of attention on the energy balance by planning agencies and acknowledgement of the vital role of energy in an expanding economy has led to a rapid increase in total output and also, more recently, to a more rational utilization of power resources. In the late 1950s there was a marked swing from coal, which had long supplied over 60 percent of the energy requirements of the USSR, to the more economical fuels, oil and natural gas. The wasteful use of wood as a fuel has steadily been decreasing, and in spite of the vast reserves of coal, oil, and gas, Soviet planners have become too conscious of these minerals as valuable raw materials for the chemical industry to go on burning them indefinitely as the main source of power. Alternative sources, such as hydroelectric power and nuclear energy, are expected to play an increasingly important role in future.

Table 1.1 shows the change in emphasis from coal to oil and gas and the increased use of hydroelectric power that had become evident by 1970. Nuclear power in 1973 accounted for only one percent of the total output of energy.[2]

The comparative share of coal, oil, gas, peat, oil shale, and wood in the total yearly output of fuel in the USSR since 1913 is shown in Table 1.2. The rise in the importance of natural gas in recent years is particularly noticeable.[3]

These changes in the comparative share of each fuel occurred simultaneously with a remarkable rise in the actual production of coal, oil, and gas, the major fuels. There has also been a rapid acceleration in the rate of growth of peat and oil shale extraction, although the growth in the use of peat has fluctuated considerably. The consumption of wood as a fuel has been lessening steadily since 1965. The

TABLE 1.1

Changes in the Soviet Energy Balance, 1913-70
(in percentage)

Form of Energy	1913	1950	1955	1960	1965	1970 (Estimated)
Coal	45.0	64.3	62.5	52.1	41.5	34.7
Oil	24.2	17.0	21.0	29.5	34.7	38.9
Gas	—	2.3	2.3	7.6	15.3	18.3
Peat	1.2	4.6	4.3	2.8	1.6	2.3
Oil shale	—	0.4	0.7	0.7	0.7	0.6
Wood	29.55	8.8	6.7	4.0	2.8	1.4
Hydroelectric energy	0.05	2.6	2.5	3.3	3.4	3.8

Sources: Energeticheskie resursy SSSR: Toplivno-energeti-cheskie resursy, Moscow 1968: 6; Tolkachev, A. S., et al., Osnovnye napravleniya nauchno-tekhnicheskogo progressa, Moscow 1971: 111.

7

TABLE 1.2

Comparative Production of Fuels in USSR, 1913-75
(in percentage of total output)

Year	Oil	Natural Gas	Coal	Peat	Oil Shale	Wood
1913	30.5	—	48.0	1.4	—	20.1
1940	18.7	1.9	59.1	5.7	0.3	14.3
1945	15.0	2.3	62.2	4.9	0.2	15.4
1950	17.4	2.3	66.1	4.8	0.4	9.0
1955	21.1	2.4	64.8	4.3	0.7	6.7
1960	30.5	7.9	53.9	2.9	0.7	4.1
1965	35.8	15.5	42.7	1.7	0.8	3.5
1970	40.4	18.9	36.1	1.7	0.7	2.2
1971	41.8	19.5	35.6	1.3	0.7	2.1
1975	44.1	23.3	29.5		3.1	

Sources: Narodnoe khozyaistvo SSSR v 1970 Godu, Moscow 1971: 183; Ekonomicheskaya gazeta no. 27, July 1972: 1; Narodnoe khozyaistvo SSSR 1922-1972, Moscow 1972: 162.

comparative growth in the output of fuels in the USSR from 1913 to 1970 is shown in Table 1.3* and Figure 1.1.

The greatest part, 88.5 percent, of the Soviet fuel reserves is coal; only 3.2 percent of the reserves is natural gas; .8 percent is peat, .7 percent is oil shale, and 6.8 percent is made up of other fuels.[4] (Precise information on Soviet oil reserves has not been published, since they came under the State Secrets Act in 1947.) It was calculated in 1971 that, in spite of increased consumption, coal reserves would suffice for another 1000 years, oil shale for 560 years, peat for 160 years, oil for about 100 years, and natural gas reserves for only 83 years (or 95 to 100 years including incidental oil-well gas).[5] Yet because of the many advantages of natural gas as a fuel, the USSR plans to increase its output to such an extent that by the end of this century it will be the principal source of energy, overtaking both coal and oil. (See Table 1.4.) It is expected that nuclear energy will then

---

*This book employs United Kingdom denominations and measures, which are the ones used in the USSR.

FIGURE 1.1

Comparative Growth in Soviet Output of Fuels, 1913-1970

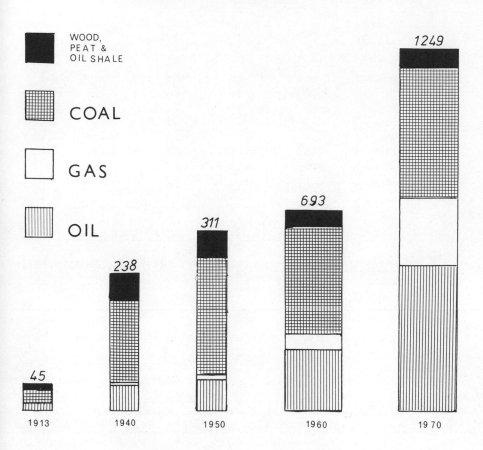

Source: Table 1.3.

TABLE 1.3

Fuel Production in USSR, 1913-71
(in millions of conventional fuel tons)

Year	Total Output	Oil	Natural and Incidental Gas	Coal	Peat	Oil Shale	Wood
1913	48.2	14.7	—	23.1	0.7	—	9.7
1940	237.7	44.5	4.4	140.5	13.6	0.6	34.1
1945	185.0	27.8	4.2	115.0	9.2	0.4	28.4
1950	311.2	54.2	7.3	205.7	14.8	1.3	27.9
1955	479.9	101.2	11.4	310.8	20.8	3.3	32.4
1960	692.8	211.4	54.4	373.1	20.4	4.8	28.7
1965	966.6	346.4	149.8	412.5	17.0	7.4	33.5
1970	1248.6	504.2	235.6	451.2	21.4	8.5	27.7
1971	1284.9	537.3	250.6	444.2	16.7	9.5	26.6

Sources: Narodnoe khozyaistvo SSSR v 1970 godu, Moscow 1971: 183; Narodnoe khozyaistvo SSSR 1922-1972, Moscow 1972: 162.

TABLE 1.4

Projected Changes in the Soviet Energy Balance, 1970-2000
(in percentage)

Form of Energy	1970	1980	1990	2000
Coal	37.1	25	23 to 21	21 to 18
Oil	39.5	38	31	26
Gas (natural and incidental)	18.4	27	30 to 32	31 to 35
Other fuels (including imported)	3.6	7	5	4
Atomic energy	0.3	2	10	16
Hydroelectric energy	1	1	1	1

Source: Melnikov, N. V., Toplivno-energeticheskie resursy SSSR, Moscow 1971: 7.

FIGURE 1.2

Soviet Energy Balance, 1950-2000

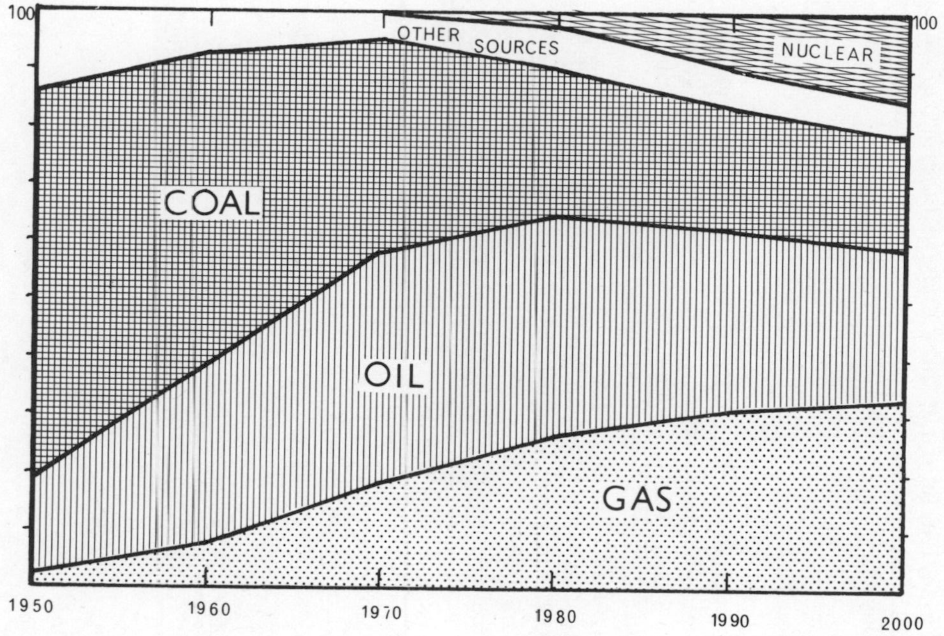

Sources: Tables 1.2 and 1.4.

11

gradually increase its contribution until it supplies most of the Soviet energy requirements.[6] The changing pattern of the Soviet energy balance in the second half of the twentieth century is shown in Figure 1.2.

In 1971 approximately 83 percent of electricity was produced by thermal stations, and this proportion is expected to be maintained in the immediate future.[7] This study will therefore concentrate on the fuel industries, but alternative sources of energy will be dealt with in Chapter 7.

NOTES

1. Stroev, K. F., et al., Ekonomicheskaya geografiya SSSR (Economic geography of the USSR), Moscow 1972: 265, Energeticheskie resursy SSSR: Toplivno-energeticheskie resursy (Energy resources of the USSR: fuel resources), Moscow 1968: 38.

2. Ibid.: 6.

3. Narodnoe khozyaistvo SSSR v 1969 godu (National economy of the USSR in 1969) Moscow 1970, 196; Ekonomicheskaya gazeta no 27, July 1972: 1.

4. Energeticheskie resursy, op. cit.: 37.

5. Melnikov, N. V., Toplivno-energeticheskie resursy SSSR (Fuel resources of the USSR), Moscow 1971: 20, 31.

6. Ibid.: 7; SSSR v tsifrakh v 1970 godu (USSR in statistics in 1970), Moscow 1971.

7. Ekonomicheskaya gazeta no 6, February 1972: 22.

The most important change in the energy balance of the USSR in recent years has been the increased importance of natural gas. While only a decade before it could be claimed that gas occupied a "relatively insignificant position in the energy economy of the Soviet Union,"[1] by 1970 gas was supplying a fifth of the country's energy requirements. It was being used to produce almost 30 percent of the electricity and 80 percent of the iron and steel and had replaced some of the alimentary raw materials formerly used in the chemical industry.[2] Over 100 million Soviet citizens were using gas as their main domestic fuel. During the eighth Five Year Plan (1966-70) gas was supplied to 1,325 towns and 34,000 villages.[3]

In several important economic regions gas had become the main source of fuel. It was 45 percent of the energy balance in the Central Economic Region, 34 percent in the Northwest, almost 50 percent in the North Caucasian, 47 percent in the Transcaucasian, and more than 55 percent in Central Asia.[4]

In the decade following the war (1946-55) the average yearly increase in output was 370 million cu.m.; in the next decade (1956-65) it reached 10,500 million, and from 1966 to 1970 it rose to 13,600 million. The rate of increase is expected to continue to accelerate from 1971 to 1975, forming an average annual rise in output of 20,000 to 24,000 million cu.m.[4] Production increased from 3,902 million in 1946 to 221,000 million in 1972, and was planned (somewhat optimistically) to reach 320,000 million in 1975.[5]

Some recent data have been added to this chapter from the BBC monitoring service, USSR Weekly Economic Reports (Oil and Gas), 1972-73.

Reasons for this sudden change in the importance of gas are
not difficult to find. Somewhat later than those in the West, Soviet
planners came to realize that gas costs less than other energy sources,
and decided to utilize the vast reserves being discovered in the 1950s.
During the Seven Year Plan (1959-65) the increased use of natural
gas in the economy brought savings of about 8 billion rubles; this was
more than double the cost of developing the gas industry during this
period.[6] Taking into account exploration, extraction, and transport,
the cost of gas in the main areas of demand is less than a third that
of coal. One reason for this is that labor productivity in extracting
gas is some 30 times higher than in the underground mining of coal.[7]
The importance of this becomes evident when one considers that the
percentage of the total production costs for electricity that is spent
on fuel has been calculated to be from 54 to 63 percent; for iron and
steel it is 23 to 24 percent; for cement it is 30 percent; and for synthetic materials it is 16 to 18 percent.[8]

It is therefore worthwhile to examine the development of the
gas industry in some detail, especially since even the more recent
works in English on the energy economy of the USSR tend to be out
of date with regard to this most rapidly expanding sector.

HISTORICAL BACKGROUND

In prerevolutionary Russia, manufactured gas was used in
Moscow, St. Petersburg, Odessa, Kharkov, Riga, and Rostov, but even
in the town centers its use was limited. By 1867 there were some
6,000 street lights burning gas in Moscow, but it was not used in
homes until much later. In 1914 there were only 2,700 flats in Moscow and 3,000 in St. Petersburg that were supplied with gas. Gas
works were small and backward in comparison with those in the West.[9]
The total amount of gas manufactured in 1913 was 17 million cu.m.,
and although some oil-well gas was produced in Azerbaidzhan and
natural gas utilized to a small extent for local needs in the Apsheron
peninsula and in the Surakhan area, there was much wastage involved.[10]
Oil-well gas was usually vented and flared.

Production increased as the industry recovered from the destruction of the Civil War, natural and oil-well gas alone supplying over
127 million cu.m. in 1924-25 and 304 million in 1928. New methods
of closed extraction cut down losses, over 500 million cu.m. being
saved in the Baku district alone in the two years 1939 and 1940, when
output reached 3,219 million cu.m.* However, in spite of a steady

---

*According to recent reports, however, losses of incidental oil-
well gas continue to cause concern.

increase in output the gas industry continued to play a minor role in the total fuel economy; in 1923 the first gas processing plant was built in Baku, but only four more were built in the period up to 1955.[11] Exploration for natural gas in Dagestan, Komi ASSR, the Ukraine, and the North Caucasus showed some success, but exploration tended to concentrate on oil-bearing regions.[12] The misleading impression that the Soviet Union lacked substantial gas resources made planners reluctant to allow a large expenditure on intensive exploration for gas fields. Up to the Second World War the natural gas fields in exploitation were concentrated in Dagestan, where 8.3 million cu.m. were extracted from 1937 to 1939. The small quantities produced at the Azov field were used in the form of bottle gas for domestic consumption and even for transport.[13] By 1940, despite rapid growth in the gas industry, only 12.6 percent (406.8 million cu.m. out of a total of 3219.1 million cu.m.) was from natural gas fields rather than from oil fields.[14]

The construction of gas pipelines had begun in a very small way before the war. In 1941 Lvov was supplied with gas from the Dashava field through a pipeline 200 mm. in diameter and 70 km. in length. In 1942 and 1943 gas pipelines were constructed from Elshanka to Saratov, and from the Buguruslanskoe and Pokhvistnevo fields to Kuibyshev, in order to supply these important industrial centers with an alternative fuel to coal and oil, which were no longer available from the areas occupied by the Germans. The Kuibyshev pipeline was 160 km. in length and 300 mm. in diameter. In June 1946 the first long-distance pipeline began to deliver gas over 788 km. to Moscow from the fields that had been discovered during the war in the Saratov oblast. By present-day standards it was very small, being only 325 mm. in diameter, but it provided valuable experience for the future construction of pipelines.[15]

The expansion of the Soviet gas industry received a definite impetus from the war, in spite of the temporary loss of fields in the Ukraine and the North Caucasus. Towards the end of 1942 the recently discovered gas field near Elshanka on the right bank of the Volga began to produce gas; in twenty days the pipeline to Saratov was completed, and industrial enterprises and power stations were rapidly converted to this new source of energy. Although pipeline capacity during the war never exceeded 1,000 million cu.m. in one year, the strategic importance of this innovation was great. Pipelines supplied Saratov and Kuibyshev with cheap fuel while relieving the overburdened railway system of many tons of solid and liquid fuels. The Sedelskoe field, southeast of Ukhta, had been discovered in 1935 (the first in the Komi ASSR), but industrial exploitation did not commence until the beginning of 1942. It was soon followed by many other fields.[16] In 1940 natural gas had provided 1.9 percent of the total fuel balance;

by 1945 it was contributing 2.3 percent, but it is worth noting that this compares rather badly with both peat (4.9 percent) and wood (14.3 percent). In 1955 the contribution of natural gas was still a mere 2.4 percent.[17]

This does not mean, however, that the industry made no progress in the decade following the war. Although only a fraction of the total investment in the oil and gas industry in both the fourth and fifth Five Year Plans was allotted to gas, many new gas fields were discovered in the Saratov, Volgograd, Orenburg, and Kuibyshev oblasts and in the Ukraine, where the large Ugerskoe field went into production in 1946. In 1948 the gas pipeline from the Dashava fields to Kiev went into operation, and by 1951 it had been extended to Bryansk and Moscow (1,330 km. in length and 500 mm. in diameter). Further important gas fields were discovered in the Northern Caucasus and in Central Asia, and in 1953 the Berezovskoe field in Tyumen oblast was found. By 1955 there were 4,860 km. of major gas pipelines in the USSR, including a double pipeline from Kokhtla-Yarve to Leningrad (length 200 km., diameter 500 mm.) and pipelines from Kokhtla-Yarve to Tallin, from Tuimazy to Ufa, from Minnibaevo to Kazan, from Archeda to Volgograd, from Tula to Moscow and others.[18]

The location of the major areas of gas extraction had also changed by 1956. In 1940 Azerbaidzhan had supplied 77.7 percent of the total amount of gas extracted in the USSR; by 1946 its share had dropped to 29.7 percent, and by 1956 it was 17.4 percent. The amount extracted in the RSFSR in 1940 was only 6.5 percent; this rose to 44.4 percent in 1946 and 47 percent in 1956. There was also an increase in the contribution of the Ukraine in those years: in 1940 it produced 15.5 percent; this rose to 25.1 percent in 1946 and 33.1 percent in 1956.[19] By 1945 over 67 percent of the gas extracted was from natural gas deposits, and the proportion of incidental oil-well gas has continued to drop.[20]

Until 1956 the amount of gas extracted each year grew very slowly. Much valuable capital was invested in costly methods of manufacturing gas from coal and oil shales and in the underground gasification of coal. The latter scheme has since been allowed to drop and seems to have been an expensive mistake. During the fifth Five Year Plan almost half the capital expenditure on the gas industry was spent in this way.

Although known reserves of natural gas increased greatly after the war, this was usually the accidental result of exploration for oil and did not immediately lead to their exploitation.

Among the many changes in the Soviet Union after the Twentieth Congress of the CPSU in 1956 was the adoption of a new fuel policy that greatly affected the development of the gas industry. In the Directives for the sixth Five Year Plan (1956-60) it was stated that the output of gas should increase by over four times, reaching 40,000 million cu.m. by 1960. In February 1957 the first conference on the further development of the gas industry was held. It dealt with such matters as intensifying exploration and exploitation of gas deposits, increasing the facilities for the storage and transport of gas, and extending the utilization of gas in the chemical industry and for domestic needs. In August 1958 a decree was promulgated on the "further development of the gas industry and the gasification of industrial enterprises and towns in the USSR." A marked increase in capital outlay brought an immediate return, in the discovery of several important new gas fields and a sharp increase in output.[21]

The new importance of the gas industry in the Soviet fuel balance has been reflected in its administrative structure. For a long time a mere subsidiary of the oil ministry, the gas industry has received increasing recognition and autonomy. In 1956 "Glavgaz USSR" was created, making the gas industry an independent branch of the economy with a single central management, and after the September Plenum of the Central Committee of the CPSU in 1965 the "Ministry of the Gas Industry of the USSR" was formed.[22]

The remarkable change in the yearly rate of growth in gas extraction that occurred in the mid-1950s can be seen from the graph for Table 2.10 (Figure 2.3). The pattern set then has continued; the yearly increase in output is now greater than the total amount of gas produced in 1955. (See Table 2.10.)

Expansion in exploratory drilling brought satisfactory results. From 1957 to 1959 several large fields were discovered in the Krasnodar krai, the Ukraine, and Central Asia, and from 1959 to 1965 exploration revealed further gas-bearing provinces in the northern Tyumen region, Tomsk oblast, and in East Turkmenia. During the Seven Year Plan alone over 230 gas and gas condensate fields were discovered.[23] The most important gas fields discovered by 1965 are given in Table 2.1.

Since 1965 important fields such as Urengoiskoe, Medvezhe, and Zapolyarnoe have been found in the northern part of Tyumen oblast.[24] The vast Vuktylskoe gas condensate field was discovered in 1966. In 1967 the Efremovskoe field in the Ukraine added its contribution to the known gas reserves and was soon followed by the Achakskoe field in Turkmenia. The Tomsk oblast and the North Caucasus have also proved rich in gas.[25]

The proportion of incidental oil-well gas in the total output of gas in the USSR has dropped considerably, in spite of the increased amounts extracted, from 1,799 million cu.m. out of 5,761 million in 1950; 7,706 million out of 45,303 million in 1960; 16,483 million out of 127,666 million in 1965; and 23,000 million out of 198,000 million in 1970. The production of manufactured gas had increased from

17

TABLE 2.1

Major Gas Fields by 1965

Gas Field	Year of First Output	Initial Reserves A + B + C (in millions of cu.m.)	Total Output by 1 January 1966 (in millions of cu.m.)
North Stavropolskoe	1956	223,400	70,900
Shebelinskoe	1956	402,300	104,100
Bilche-Volitskoe	1949	41,000	14,800
Ugerskoe	1946	36,800	25,000
Rudkovskoe	1957	32,100	13,500
Maikopskoe	1960	94,900	10,700
Berezanskoe	1963	61,000	9,600
Leningradskoe	1958	57,600	14,900
Staro-Minskoe	1961	33,600	8,600
Stepnovskoe	1958	27,600	17,600
Korobkovskoe	1961	89,700	11,600
Gazli	1961	480,000	25,700
Karadag	1956	34,100	23,900

Source: Energeticheskie resursy SSSR: Toplivno-energetiches-kie resursy, Moscow 1968: 457.

429 million cu.m. in 1950 to 1,613 million in 1956 and to 1,911 million by 1960; it has since dropped to a steady 1,700 million cu.m.; as a proportion of all gases produced it has sharply decreased in importance each year.

Over the recent years the changes in the gas industry have been so great that it would be sensible to consider them separately. There are four main stages in the process of exploiting gas and other fuel resources: (1) exploration and location of reserves; (2) production; (3) transportation and storage; and (4) utilization in the economy.

PRESENT AND FUTURE DEVELOPMENT

Exploration and Location of Reserves

In discussing Soviet mineral resources it would seem advisable to retain the accepted Soviet classification system, since there is no exact equivalent in Western terminology. Mineral resources are divided into four main categories, A, B, C_1 and C_2, according to the extent to which they have been explored.

Category A gas reserves have been fully explored in areas outlined by productive wells and reliably established by experimental exploitation.

Category B describes gas reserves in areas where drilling has given favorable indications of commercial gas possibilities, borne out by a commercial flow from at least two wells.

Category C_1 defines reserves in locations that geological and geophysical data show to be favorable to the accumulation of gas. Porosity and permeability have been established by drilling or by analogy with neighboring deposits that have already been explored. A commercial flow of gas must be obtained from at least one well in the estimated area.

Category C_2 relates to gas in new structures of gas-bearing provinces with strata of a type that are known to be productive in other deposits. It also relates to reserves of known fields situated in unexplored tectonic blocks and strata that favourable geological and geophysical data suggest are likely to prove productive.

Two further categories exist for gas and oil reserves known as "predicted" (prognoznye) reserves. Category D_1 covers reserves in horizons that have already given indications of being productive and that may be assumed from geological prospecting to contain certain quantities of gas.

Finally, Category D_2 refers to gas in possibly productive horizons that have shown no definite indications as yet because of insufficient prospecting.[26]

While there is no doubt that the Soviet Union contains a very high proportion of the world's natural gas reserves, it is impossible to state the precise percentage, since classification of reserves differs so greatly. The nearest comparison would be that of reserves that have been discovered mainly by exploratory drillings; in other words, the sum of Soviet categories A, B, C_1 and C_2 can be compared to the sum of "proven" plus "possible" reserves in other parts of the world.

A study of world reserve figures for July 1970 reached the following conclusions. Out of a total of proved reserves of 46,676,000 million cu.m., the USSR came first with 32 percent (14,983,000 million cu.m.) and the United States was second with 16.7 percent (7,793,000 million cu.m.). Iran, with 12.8 percent, is the only other country with comparable reserves. Since the other Eastern European countries combined have a further 1.4 percent (637 billion cu.m.), the Comecon bloc is assured of a solid base for expanding its output of natural gas. While the actual figures for the world's reserves vary greatly, the share allotted to the USSR is usually about 30 percent, and this is actually rising with the results of present exploration.[27]

However, since the USSR claims to have 11,135,200 sq.km. of favorable gas-bearing areas, with a further 4 million sq.km. on the continental shelf, compared to 4,630,000 sq.km. in the United States, its reserves must be less densely contained, and a large proportion is certainly situated in inaccessible areas with extremely hostile climatic conditions.[28]

Potential reserves (the sum of all categories) in 1966 were estimated at 67×10^{12} cu.m. Following the discovery of huge gas fields in Western Siberia, Komi ASSR, and Orenburg oblast, figures for January 1971 were calculated at nearer 83×10^{12} cu.m., while radio broadcasts in May 1973 put potential reserves at around 100×10^{12} cu.m.

This yearly expansion of reserves has been the result of constant exploratory drilling, over 87 million meters having been drilled between 1920 and 1969. The amount of exploratory drilling that has been undertaken naturally varies greatly from area to area, depending usually on when each area was first discovered. The average in Azerbaidzhan was 272 m. per sq.km., but in Kuibyshev oblast it was 92.5 m. per sq.km. and in Siberia only 1.1 m. per sq.km.[29]

In 1955 only 235,000 meters were drilled in search of gas, but by 1959, with the initiation of separate prospecting for gas throughout the USSR, this figure rose to 918,000 meters and has since continued to rise. From 1955 to 1970 prospectors drilled over 35 million meters, expanding $A + B + C_1$ reserves to over 13×10^{12} cu.m. Exploratory drilling was thus very rewarding, with every meter that was drilled increasing known reserves by an average of 577,000 cu.m.[30] This success rate, which was made possible by the vast discoveries in northern Tyumen, conceals the less productive drilling figures for the early 1960s, when the actual increments in reserves fell far short of plan. In many regions in the European part of the USSR, in the North Caucasus, the Lower Volga area, Bashkir ASSR, and the Ukraine, various difficulties arose in calculating the possible reserves, since the geological situations in which the gas was located were more complex than had been expected.[31]

In the early 1960s reserves contained in the larger fields, Shebelinskoe, Stavropolskoe, and Gazli, ranged from 300,000 to 500,000 million cu.m. Reserves in each of the more recently discovered fields, Zapolyarnoe, Medvezhe, Vuktylskoe, Orenburgskoe, and Shekhitlinskoe, are as great as 1.5×10^{12} to 2×10^{12} cu.m., and in Urengoiskoe field they are over 3×10^{12} cu.m. These recent discoveries have naturally raised the economic viability of gas prospecting in the decade since 1960. For every meter drilled from 1961 to 1965 an average of 202,000 cu.m. was added to reserves at a cost of 1.06

rubles per 1,000 cu.m.; from 1960 to 1970 every meter drilled brought 1,058,000 cu.m. at a cost of only .22 rubles per 1,000 cu.m. of reserves.[32]

In northwestern Siberia prospectors have had to contend with subarctic conditions. Tundra, lakes, flooded rivers, and permafrost from .5 meters to 100 meters in depth make drilling a much more complicated operation than in more temperate climates. The temperature of the surface soil at Messoyakha, for example, can vary from +45° C to -58° C. With average temperatures in winter from -15° to -20° C and wind speeds averaging about 11 meters per second (25 mph), even supplying regular meals can be a problem.[33]

The expansion of Soviet gas reserves since 1957 is shown in Table 2.2.

TABLE 2.2

Expansion of Natural Gas Reserves
$(A + B + C_1)$, 1957-75
(in millions of cu.m.)

Year (January)	Reserves
1957	862,300
1958	1,095,600
1959	1,584,800
1960	2,202,400
1961	2,336,100
1962	2,547,400
1963	2,786,500
1964	3,061,600
1965	3,219,700
1966	3,565,900
1967	4,431,700
1968	7,752,500
1969	9,470,000
1970	12,091,800
1971	15,500,000
1972	18,000,000
1975	24,400,000 (planned)

Sources: Lvov, M. S. Resursy prirodnogo gaza SSSR (Moscow 1969), 100; Gazovaya promyshlennost No. 1, January 1970, 4; No. 4 April 1970, 10; No. 3, March 1971, 1; No. 3, March 1972, 46.

FIGURE 2.1

Expansion of Natural Gas Reserves, 1957–1972

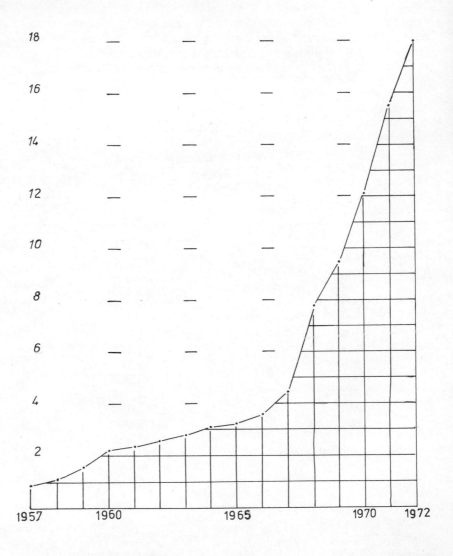

Source: Table 2.2.

As may be seen more clearly from the graph based on these figures (Figure 2.1), the most rapid expansion of reserves took place in the late 1960s and early 1970s.

Together with this expansion of total reserves it is also worth noting the change in regional distribution of the reserves. As can be seen in Table 2.3, the percentage of reserves in European areas has dropped sharply since 1956. By 1972 a full 80 percent of the gas reserves in the USSR was beyond the Urals. The percentage contribution of Eastern Siberia, the Far East, Central Asia, and Western Kazakhstan rose rapidly until 1966, when it also began to drop. The main reason for these changes has been the jump in the share of the Tyumen oblast, which in 1970 had 57 percent of all Soviet reserves, rising to 70 percent by 1973.[34]

The most important gas fields (with reserves of over 100,000 million cu.m.) that were discovered between 1965 and 1970 are given in Table 2.4. By 1970 the Vuktylskoe, Efremovskoe, and Achakskoe fields had already been put into exploitation, and all of them are likely to play a vital part in the Soviet fuel economy in the near future.[35] The whole process has been accelerated to such an extent that some fields discovered since 1970 were already being exploited by 1972. The huge Naipskoe field in Turkmenia, for example, was producing over 3 million cu.m. per day by June 1972.

The locations of these vast reserves and their distances from the centres of consumption is naturally of great importance. The areas with the greatest density of gas reserves are shown on Map 2.1, and their quantitative distribution is given in Table 2.5.

The Volga-Ural oil and gas province stretches from the White Sea in the north, includes the Ukhta and Pechora areas, and merges with the Emba gas-bearing region in the south. Gas fields have been explored in the Komi, Tatar, and Bashkir autonomous republics and in the Kuibyshev, Orenburg, Perm, Saratov, Volgograd, and Astrakhan oblasts. Gas fields in this province are located mainly in the Devonian and Carboniferous strata and have the advantage of being in horizons comparatively near the surface (1,500 to 2,500 m.). Among the most important gas fields of this province are the Korobkovskoe (250 km. north of Volgograd), Stepnovskoe, and Uritskoe (in the Saratov oblast).[36]

The Vuktylskoe gas condensate field in the Komi ASSR and the Orenburgskoe gas condensate field are outstanding recent discoveries; in fact from 1969 to 1970 the greatest expansion for this province was in Komi ASSR (150,000 million cu.m.) and Orenburg oblast (300,000 million cu.m.). By 1972, with reserves of 1,600,000 million cu.m., the Orenburgskoe was by far the largest field in the European part of the USSR. There has also been some increase in the reserves in the other areas except for Saratov, where expansion of reserves has not matched output.[37]

23

TABLE 2.3

Changes in the Distribution of Natural Gas Reserves, 1956-70

I

(in millions of cu.m.)

Area	1956	1961	1966	1970
USSR	691,900	2,336,100	3,565,900	12,091,800
European USSR	642,200	1,591,000	1,802,500	2,628,100
Western Siberia	4,100	50,200	455,100	7,087,900
Of which Tyumen oblast	4,100	50,200	400,800	6,853,800
Eastern Siberia and the Far East	4,800	32,600	146,800	361,100
of which Yakut ASSR	–	20,800	77,500	276,700
Central Asia and Western Kazakhstan	40,800	662,300	1,161,500	2,014,700

II

(in percentage)

Area	1956	1961	1966	1970
USSR	100	100	100	100
European USSR	92.8	68.2	50.6	21.8
Western Siberia	0.6	2.1	12.8	58.6
Of which Tyumen oblast	0.6	2.1	11.2	57.0
Eastern Siberia and the Far East	0.7	1.4	4.1	2.9
Of which Yakut ASSR	—	0.9	2.2	2.3
Central Asia and Western Kazakhstan	5.9	28.3	32.5	16.7

Sources: Gazovaya promyshlennost no. 4, April 1970: 10; Pravda, 17 March 1972: 3.

TABLE 2.4

Most Important Gas Fields
Discovered Between 1965 and 1970
(in millions of cubic meters)

Gas Fields	Reserves as of January 1970	
	$A + B + C_1$	$A + B + C_1 + C_2$
Orenburgskoe	504,000	900,000
Vuktylskoe	374,000	500,000
Zapolyarnoe	1,570,000	1,620,000
Gubkinskoe	352,600	353,000
Urengoiskoe	2,431,000	7,500,000
Medvezhe	1,548,000	1,760,000
Arkticheskoe	5,000	180,000
Vyngapurovskoe	10,000	200,000
Myldzhinskoe	91,600	100,000
Sredne-Vilyuiskoe	189,900	250,000
Maastakhskoe	32,000	180,000
Efremovskoe	85,200	130,000
Kandymskoe	35,000	108,000
Urtabulakskoe	32,000	153,000
Achakskoe	160,000	165,000
Saman-Tepenskoe	105,000	120,000
Beurdeshikskoe	—	104,000
Shekhitlinskoe	162,000	400,000
Pilyatkinskoe	45,000	200,000
North Komsomolskoe	50,000	600,000
Yubileinoe (Tyumen oblast)	50,000	2,000,000
South Russkoe	—	500,000
Komsomolskoe	378,000	380,000
Soleninskoe	6,000	150,000

Source: Gazovaya promyshlennost no. 4, April 1970: 11.

The Caucasus oil and gas province covers five main gas-bearing areas: Krasnodar krai, Stavropol krai, Dagestan ASSR, Checheno-Ingush ASSR, and Azerbaidzhan SSR. The first two named have the greatest reserves, and although output is now so high that there was a drop in reserves (1969-70), extraction of gas is guaranteed for several years to come. The most important gas fields of Krasnodar krai are Maikopskoe (situated 15 km. north of the town of Maikop),

Leningradskoe (180 km. northwest of Krasnodar), Staro-Minskoe (160 km. north of Krasnodar), Berezanskoe, and Kanevskoe. In Stavropol krai the North-Stavropolskoe is the largest gas field.[38] In Azerbaidzhan, gas is to be found in the east of the main Caucasian range, in part of the Kura lowlands, in the Apsheron peninsula, and under the adjacent shelf of the Caspian Sea. The main gas field in Azerbaidzhan is the Karadagskoe.[39] The largest expansion of reserves from 1969 to 1970 was in the Dagestan ASSR, where they increased by over 25 percent (11,000 million cu.m.).

In Siberia there are two vast oil and gas provinces, the West Siberian, stretching from the Ob to the Enisei, and the East Siberian, around the Lena and Vilyui rivers.

There have been striking recent developments in West Siberia. In 1967 the Messoyakhskoe field was discovered near the mouth of the Enisei. Further discoveries in 1969 brought the reserves of this area to almost 400,000 million cu.m. Norilsk is supplied by pipeline from this source. Around the Berezovskoe area over 21 fields have been discovered, with total reserves of 200,000 million cu.m. The biggest is the Punginskoe, with reserves of 60,000 million cu.m., the output of which is being directed to the industries of the Urals. Discoveries in the Tomsk oblast, including the Myldzhinskoe field, have brought reserves there for supplying gas to Tomsk, Novosibirsk, and Kemerovo to 233,000 million cu.m. Other important fields are the Arkticheskoe and Novo-Portovskoe; Medvezhe, Yamburgskoe, and Urengoiskoe (which are among the biggest fields ever known); Tazovskoe, Zapolyarnoe, and Russkoe; and Gubkinskoe and Komsomolskoe. In January 1970 the reserves in Tyumen oblast were put at 6,773,800 million cu.m. and those in Tomsk oblast at 233,500 million cu.m. By 1973, West Siberian known reserves had increased to over 12,000,000 million cu.m.[40]

East Siberia has some 4,157,000 sq.km. of gas-bearing regions, about 27.7 percent of the USSR total, but as yet has not been explored intensively. Among the most thoroughly explored fields are the Ust-Vilyuiskoe, supplying Yakutsk and Pokrovsk with gas; the Sredne-Vilyuiskoe (which is the largest in East Siberia); and Maastakhskoe gas condensate fields. In 1970 the Yakut ASSR had reserves of over 276,700 million cu.m, but by 1972 predicted reserves were as high as 12,800,000 million.[41]

In the Ukrainian SSR there are three rich gas-bearing areas. To the west, large gas accumulations have been explored in the eastern Carpathians, where the Rudkovskoe, Ugerskoe, and Bilche-Volitskoe gas fields are situated. Further deposits stretch from the Eastern Ukraine, to north of the Dnepr river in Belorussia. The Shebelinskoe and Efremovskoe fields have been the most important producers of natural gas in recent years. The third gas-bearing area of the Ukraine

TABLE 2.5

Distribution of Natural Gas Reserves $(A + B + C_1)$

Area	Gas-Bearing Areas in sq. km.	Exploratory Drilling, m. per sq. km., 1920–69
USSR	11,135,200	7.8
European USSR	1,365,600	43.0
RSFSR	8,765,500	6.4
Komi ASSR	254,600	7.8
Bashkir ASSR	95,300	80.2
Perm oblast	118,700	25.6
Kuibyshev oblast	53,700	92.5
Orenburg oblast	91,900	31.3
Saratov oblast	100,200	25.8
Volgograd oblast	88,400	30.8
Krasnodar krai	82,400	68.2
Stavropol krai	70,500	38.0
Checheno-Ingush ASSR	16,000	289.0
Dagestan ASSR	28,000	60.0
Sakhalin oblast	62,600	36.3
Siberia	5,477,700	1.1
Other areas of the RSFSR	2,225,500	3.1
Ukrainian SSR	283,900	28.0
Azerbaidzhan SSR	36,600	272.0
Kazakh SSR	1,119,000	3.3
Central Asia	715,700	11.2
Uzbek SSR	201,000	16.8
Turkmen SSR	468,200	7.3
Tadzhik SSR	29,900	18.0
Kirghiz SSR	16,600	43.7
Georgian SSR	34,500	19.8
Armenian SSR	10,900	9.8
Other areas of the USSR	169,100	8.3

Reserves as of January 1969 in millions of cu.m.	Increase in Reserves in 1969 in millions of cu.m.	Reserves as of January 1970 in millions of cu.m.
9,470,300	2,784,500	12,091,800
2,165,200	578,000	2,628,100
6,917,800	2,448,000	9,296,000
238,100	150,000	382,600
53,000	2,000	54,400
38,800	2,000	40,000
5,200	1,000	6,100
238,100	300,000	537,400
75,100	2,000	73,000
84,500	7,000	87,600
390,100	2,000	364,700
230,900	8,000	223,000
8,300	—	8,300
36,700	11,000	46,100
66,800	5,000	71,500
5,431,600	1,955,000	7,377,500
21,400	3,000	23,800
694,200	84,000	725,900
51,600	5,000	54,200
182,000	2,000	182,900
1,624,700	244,500	1,831,800
712,300	47,000	730,100
864,800	195,000	1,052,500
31,600	1,500	32,500
16,000	1,000	16,700
—	1,000	1,000
—	—	—
—	—	—

Source: Gazovaya promyshlennost no. 4, April 1970: 10.

Legend

A   Carpathian

B   Dnepr

C   Crimean

D   Caucasus

E   Volga–Ural

F   Emba

G   Central Asian

H   West Siberian

I   East Siberian

J   Sakhalin

MAP 2.1

Natural Gas Reserves

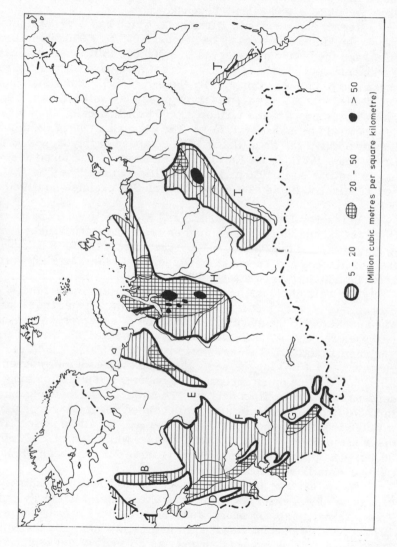

Source: Gazovaya promyshlennost no. 4, April 1970.

is in the Crimea, in the north of which are the Dzhankoiskoe and Glebovskoe fields. In 1970 the reserves of gas in the Ukrainian SSR were over 725,000 million cu.m.[42]

In Central Asia the main reserves are in the Turkmen SSR, which had 1,052,500 million cu.m. in 1970, and the Uzbek SSR, which had 730,100 million cu.m. Among the largest explored fields in Turkmenia are the Darvazinskoe, Bairan-Aliiskoe, Naipskoe, Shekhitlinskoe, Achakskoe, and Saman-Tepenskoe.

Recent discoveries in Turkmenia, which has over 30 fields, raised the known reserves as of October 1972 to 1,900,000 million cu.m. and the predicted reserves to 9,000,000 million cu.m., making Turkmenia second only to Tyumen oblast as a base for the future expansion of the gas industry. By the end of 1972 Uzbekistan had 46 gas fields, with $A + B + C_1$ reserves of 829,300 million cu.m., to which it was hoped that a further 306,000 million cu.m. known reserves would be added by 1975.[43] The giant Gazlinskoe field in the Bukhara-Khiva gas area (Uzbek SSR) has been supplying gas to the Urals since 1961. In the Kazakh SSR there are some large deposits of gas, estimated in 1970 at 182,900 million cu.m. The Zhetybaiskoe and Tenginskoe fields east of the Caspian Sea promise substantial output.[44]

In the Far East both Sakhalin and Kamchatka seem to be promising areas, and there may be further reserves around the Amur river, Khabarovsk, and the Pacific coast. The largest fields in Sakhalin are the Volchinskoe and Tungorskoe. Reserves in the Far East were about 83,200 million cu.m. in 1970.[45]

The role of ethane, propane, and condensate in the Soviet economy is of sufficient importance to warrant a separate mention when considering the quantity and location of reserves. As of 1969, reserves rated $A + B + C_1$ were mainly in the RSFSR, the Ukraine, Turkmenia, and Uzbekistan, as may be seen in Table 2.6. Predicted reserves $(D_1 + D_2)$ for the USSR as of 1969 were as follows: ethane, 2,380,000 million cu.m.; propane and butane, 3,190 million tons; condensate, 4,860 million tons.[46] These were distributed as shown in Table 2.7. When the location of predicted reserves is compared with the location of the $A + B + C_1$ reserves, it will be noticed that the share of East Siberia is much greater, while that of West Siberia drops. The proportion of predicted reserves in European USSR and Central Asia is still considerable.

Soviet experts are generally optimistic about the continued expansion of gas reserves in the future, especially in the Tyumen, Komi, Arkhangelsk, Orenburg, and Enisei estuary areas, where by 1973 several important discoveries had been made. Expansion up to the year 2000 is expected, mainly as a result of the exploration of the less intensively drilled areas of the USSR and also of the

32

TABLE 2.6

Distribution of Reserves of Ethane, Propane and Butane, and Condensate (A + B + C$_1$)

Area	Ethane (in millions of cu.m.)	Propane and Butane (in millions of tons)	Condensate (in millions of tons)
USSR	190,000	262	183
RSFSR	110,000	122	119
European RSFSR	64,000	83	79
Western Siberia	33,000	26	19
Eastern Siberia	12,000	11	19
Far East	1,000	2	2
Ukrainian SSR	22,000	86	28
Turkmenian SSR	23,000	18	17
Uzbek SSR	24,000	20	10
Kirgiz SSR	400	0.4	0.4
Tadzhik SSR	1,600	0.9	0.5
Kazakh SSR	8,000	12.7	3.1
Azerbaidzhan SSR	1,000	2	5

Source: Gazovaya promyshlennost no. 8, August 1969.

continental shelf. The central areas of the European USSR, the Tunguska basin, the northern regions of the East Siberian gas province, the northeastern USSR, Kamchatka and the Far East, and parts of the Kazakh and Kirgiz republics seem to be the most promising land areas for expanding predicted reserves (D$_1$ + D$_2$).

It has been calculated that there are some 4 million sq. km. of favorable gas-bearing areas in the most accessible parts of the continental shelf. Exploration of the seabeds is expanding. In June 1972 an expedition was mounted by the Institute of Arctic Geology to continue the search for oil and gas in the Arctic, principally in the Barents Sea.[47]

While there is too much guesswork involved for the results of long-range forecasts to be completely reliable, one expert has claimed that the predicted (D$_1$ + D$_2$) reserves of natural gas contained in the continental shelf around the USSR may have reached the figures given in Table 2.8 by the year 2000.[48]

TABLE 2.7

Distribution of Predicted Reserves of Ethane, Butane
and Propane, and Condensate $(D_1 + D_2)$
(in percentage)

Area	Ethane	Propane and Butane	Condensate
European USSR	35.2	45.3	46.0
Western Siberia	5.9	8.1	6.1
Eastern Siberia	34.1	26.0	36.0
Far East	0.4	0.7	0.4
Central Asia and Kazakh SSR	22.2	18.0	6.0
Transcaucasia	2.2	2.9	5.5

Source: Gazovaya promyshlennost no. 8, August 1969.

TABLE 2.8

Possible Seabed Reserves of Natural Gas
$(D_1 + D_2)$ by the Year 2000

Sea	Reserves (in millions of cu.m.)
Baltic	800,000
Barents (Archangel oblast)	3,000,000
Kara (Tyumen oblast)	11,000,000
Okhotsk (Sakhalin)	320,000
Azov (Krasnodar krai)	1,000,000
Black (Ukrainian SSR)	1,000,000
Caspian	5,600,000
(Azerbaidzhan SSR)	2,500,000
(Turkmen SSR)	1,500,000

Source: Gazovaya promyshlennost no. 1, January 1970: 5.

TABLE 2.9

Possible Expansion of Potential Gas Reserves
$(A + B + C_1 + C_2 + D_1 + D_2)$ by the Year 2000
(in thousands of millions of cu.m.)

Area	1958	1966	1971	2000 (estimated)
USSR	20,400	67,300	83,000	150,000
European USSR	7,500	16,300	20,600	32,500
East Siberia	2,570	16,900	16,900	23,500
West Siberia	3,640	16,600	28,000	44,400
Far East	40	700	700	1,200
Central Asia and Kazakh SSR	5,870	14,700	14,700	22,700
Transcaucasia	800	2,100	2,100	3,700
Seabed	—	—	—	22,000

Source: Gazovaya promyshlennost no. 1, January 1970: 6.

It is also possible, as Soviet experts claim, that by 1980 drilling technology will have advanced enough to allow the intensive exploitation
of reserves in layers at depths greater than 5,000 meters. Making
allowance for such progress, a figure as high as 150×10^{12} cu.m.
has been suggested for the possible potential reserves in the USSR
by the year 2000. The picture this would give of the expansion of
reserves $(A + B + C_1 + C_2 + D_1 + D_2)$ by then is shown in Table 2.9.
The average yearly expansion of potential reserves from 1958 to 1966
was 5,840,000 million cu.m., while the comparative figure for 1960
to 1971 was 3,140,000 million cu.m. The estimate of 150×10^{12} cu.m.
for the year 2000 allows for a further drop in the yearly average
(1971-2000) to 2,230,000 million cu.m. If, however, one includes in
these calculations the probability of even more rapid technological
progress in future decades than in the immediate past, and if one
takes into account the comparatively simple geological structure of
these favorable areas, then efficiency rates in exploratory drilling
should be fairly high, about 300-340 thousand cubic meters added per
meter drilled. To allow for a rise in the yearly consumption of gas
to a possible 2×10^{12} cu.m. by the year 2000, it will be necessary to
increase the level of exploratory drilling to a yearly average of 7 to
8 million meters by 1990-2000.[49]

This large-scale activity would be concentrated in the regions with large enough supplies of gas to feed high-capacity pipelines for several years. Among these could be the Komi ASSR and north Tyumen oblast and their sea areas, the Enisei and Lena estuary areas, Yakut ASSR, and parts of Central Asia. It is also considered possible that reserves in Azerbaidzhan and the Caspian will prove large enough to permit the yearly output there to reach over 50,000 million cu.m. by the end of the century.[50]

It can therefore be claimed that natural gas reserves in the USSR as known at present are sufficient to allow the rapid increase in output of the past decade to continue for the next 50 years. Recent discoveries would indicate that even with a much increased yearly extraction rate these reserves will not have been exhausted by the end of the century. On the contrary it is quite possible that further exploration will have expanded commercial reserves to an even higher figure than at present.

A study of the location of these reserves reveals, however, that some important industrial regions, such as the Center, Belorussia, and the Baltic republics, are almost totally without local supplies of natural gas, while there are vast reserves in areas so far north as to be virtually uninhabitable. This problem can only be resolved by a further expansion of the gas pipeline network.

Production

As storage of natural gas after production is both complicated and costly, the rational operation of the gas industry calls for a precise relationship among its three sectors, extraction, transmission, and consumption. It is also important to economize on men and equipment by lowering the total number of gas wells while increasing the individual output of those in use. Extraction policy in the USSR has therefore favored developing fields in the vicinity of industrial areas first, even when their reserves are comparatively limited, and next developing those fields in more remote areas that are large enough to justify the construction of long-distance pipelines to link them with the points of consumption.

Gas extraction after the war was from comparatively small deposits; the Amanakskoe field had reserves of 153 million cu.m.; the Buguruslanskoe had 35 million cu.m.; and the biggest, the Kurdyumo-Elshanskoe, had 14,000 million cu.m. Yet to the nearby industries of Saratov, Orenburg and Kuibyshev this natural gas was valuable enough to ensure the continued expansion of gas in the national fuel balance.[51] It is now claimed that, until the advent of cheap nuclear power as the main source of energy, natural gas should be the largest contributor to the expansion of the Soviet energy balance, since it is at present the most economical. For every thousand rubles invested in the coal and oil industries there is a return of 48 tons of conventional fuel; the gas industry, however, gives 296 tons. The return per worker in the coal industry is 38 tons; in the oil industry it is 330 tons; but the equivalent figure for the gas industry is 2,100 tons.[52]

In 1957 the average daily output per well was 81,000 cu.m.; in 1970 the average was up to 180,000 cu.m., with Shebelinskoe and Gazli producing over 600,000 cu.m.[53] By 1971 two wells at Urengoi and Medvezhe were each supplying over 2 million cu.m. per day.[54] The number of wells in operation in 1961 was 1,200, rising to over 3,000 by 1969, but because of the increase in the total gas extraction figures over the same period it is still possible to claim a significant improvement in the average output per well.[55]

The extraction of natural gas in the USSR has increased very rapidly since the mid-1950s and must continue to accelerate in the future. A representative of the USSR Oil Research and Geological Prospecting Institute predicted during a radio broadcast in April 1972 that by the year 2000 the Soviet Union will have to be producing at least 1,500,000 million cu.m. a year. Progress in this respect is shown in Table 2.10 and Figure 2.2.

The USSR is now second only to the United States in the production of natural gas. In 1971, when world output was 1,110,000 million cu.m., the major producers were the United States with 624,000 million, the USSR with 212,000 million, Canada with 74,000 million, the Netherlands with 44,000 million, Rumania with 26,000 million, Britain with 19,000 million, Mexico with 18,000 million, and Iran with 16,000 million.

Soviet political and economic journalists show a regrettable prediction for illustrating comparative rates of growth in branches of Soviet and American industry by the use of graphs based on percentage increases. A more balanced view of the progress made by both countries in natural gas production in the 1960s is given by the comparative production figures in Table 2.11 and Figure 2.3.

The yearly production of natural gas is much greater in the United States than in the USSR, and these figures show that the average quantitative increase each year has also been higher in the United States. This would imply that this gap between the two powers is not likely to be closed for several years to come. The Soviet Union, however, is in a much stronger position with regard to the ratio of consumption to reserves. In 1972 the United States had sufficient proven reserves for only ten years, while the USSR could continue to extract gas at the 1972 level for about fifty years.[56]

Production of gas in the eighth Five Year Plan fell far short of expectations. In 1961 it was hoped that by 1970 output would reach

TABLE 2.10

Extraction of Natural Gas (Including Oil-Well Gas)
in the USSR, 1928-2000
(in millions of cubic meters)

Year	Output	Year	Output
1928	304	1952	6,384
1929	330	1953	6,869
1930	520	1954	7,512
1931	845	1955	8,981
1932	1,049	1956	12,067
1933	1,063	1957	18,583
1934	1,531	1958	28,085
1935	1,806	1959	35,391
1936	2,050	1960	45,303
1937	2,179	1961	53,981
1938	2,208	1962	73,525
1939	2,531	1963	89,832
1940	3,219	1964	108,550
1941	3,463	1965	127,666
1942	2,045	1966	142,962
1943	1,852	1967	157,445
1944	2,405	1968	169,108
1945	3,278	1969	181,102
1946	3,902	1970	197,946
1947	4,830	1971	212,000
1948	5,219	1972	221,000
1949	5,396	1973[a]	236,000
1950	5,761	1974[a,b]	260,000
1951	6,252	1975[b]	320,000
		2000[b]	1,500,000

[a]Original target for 1973 was 250,000 and for 1974 was 280,000.
[b]Planned as of February 1974.

Sources: Lvov, M. S., Resursy prirodnogo gaza SSSR, Moscow 1969: 25; "Pokazateli razvitiya gazovoi promyshlennosti SSSR," Table 5, in Gazovaya promyshlennost no. 4, April 1970; Gazovaya promyshlennost no. 2, February 1971: 43; Ekonomicheskaya gazeta no.5, January 1972: 3; Pravda, 21 December 1971: 1; Pravda, 26 January 1974: 2.

Figure 2.2

Expansion of Natural Gas Production in USSR, 1945-1973

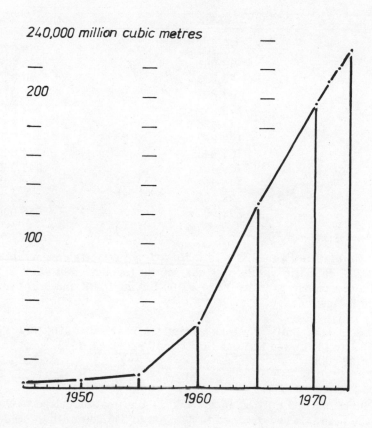

240,000 million cubic metres

200

100

1950   1960   1970

Source: Table 2.10.

TABLE 2.11

Comparative Production of Natural Gas
in USSR and United States, 1960-71
(in millions of cu.m.)

Year	USSR	United States
1960	45,303	359,673
1961	53,981	373,276
1962	73,529	390,810
1963	89,832	415,313
1964	108,550	437,842
1965	127,666	454,198
1966	142,962	487,240
1967	157,445	514,557
1968	169,108	547,152
1969	181,102	586,112
1970	198,000	625,000
1971	212,000	635,000

Note: The figures in this table are not strictly comparable, since the calorific values in k.cal./cu.m. (measured at standard sea level atmospheric pressure) are 9,500 for the USSR and 9,211 for the United States.

Sources: United Nations Statistical Yearbook 1970, New York 1971: 210-11; United Nations Statistical Yearbook 1972, New York 1973; 182-183.

310,000 to 325,000 million cu.m., but in the final directives for the 1966-70 plan this had dropped to 225,000 to 240,000 million, and in actual fact the amount produced in 1970 was under 200,000 million. Similarly, in the ninth Five Year Plan it was originally planned to produce 230,000 million cu.m. in 1972, but only 221,000 million were actually produced. In 1973 the original plan of 250,000 million was altered to 238,000 million,[57] and only 236,000 million were in fact extracted.

In 1970 the administration of the gas industry was simplified, with the central ministry relegating responsibility through large associations (obedinenie) to the enterprises (predpriyatie). At the beginning of the 1970s the main associations contributed to the total USSR output of natural gas as shown in Table 2.12. The individual shares of the other associations were relatively small.

FIGURE 2.3

Comparison of Natural Gas Production in USA and USSR

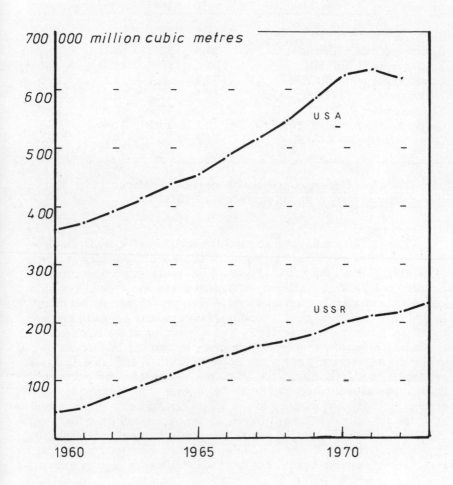

Source: Table 2.11.

TABLE 2.12

Output of Natural Gas, by Association, 1970-73
(in thousands of millions of cu.m.)

Association	1970	1971	1972	1973 (planned)
Ukrgazprom (Ukraine)	55.0	57.1	58.8	58.0
Bukharaneftegaz (Uzbekistan)	31.4	32.6	33.1	36.1
Kubangazprom (RSFSR)	22.5	20.6	14.3	9.6
Stavropolgazprom (RSFSR)	15.7	15.4	15.2	14.0
Turkmengazprom (Turkmenistan)	11.8	15.1	19.8	29.0
Komigazprom (RSFSR)	6.2	10.1	12.8	15.0
Glavtyumengazprom RSFSR)	9.2	n.a.	11.4	18.3
Orenburggazprom (RSFSR)	0.8	n.a.	4.0	4.2

Sources: Gazovaya promyshlennost no. 3, March 1971: 2; Ekonomicheskaya gazeta no. 8, February 1973: 2.

The extraction of natural gas (without oil-well gas) in the producer republics since 1958 is shown in Table 2.13. As may be seen more clearly from the graph (Figure 2.4), the main producer has remained the RSFSR; the highest extraction areas for natural gas have been Krasnodar krai, Stavropol krai, Volgograd oblast, and Saratov oblast. The contribution of Tyumen oblast has only begun to reflect its share of reserves since 1967. The proportion of the total extracted in these areas up to 1970 is given in Table 2.14. Krasnodar krai ("Kubangazprom") has since shown a marked drop in output, and production is also falling in the Stavropol, Volgograd, and Saratov fields. The annual increase for the USSR as a whole has made the proportion extracted in these areas drop even further. In Stavropol krai, for example, 79,000 million cu.m. was produced from 1965 to 1970, but in the same period proved reserves dropped to 50,000 million cu.m. Drilling deeper than 3,500 meters may reveal sufficient reserves to maintain output, but local specialists are not optimistic. The importance, however, of Tyumen oblast as a producer of natural gas is certain to continue to grow in the near future. In 1972 it produced 11,600 million cu.m. and should be able to reach an annual output of 500,000 million cu.m., since over 70 percent of the known gas reserves of the USSR are situated in Tyumen oblast. The 1973 output was expected to be over 18,000 million cu.m. By 1975 the Orenburg field is expected to contribute about 30,000 million cu.m. a year,

FIGURE 2.4

Production of Natural Gas by Republics, 1958–1970

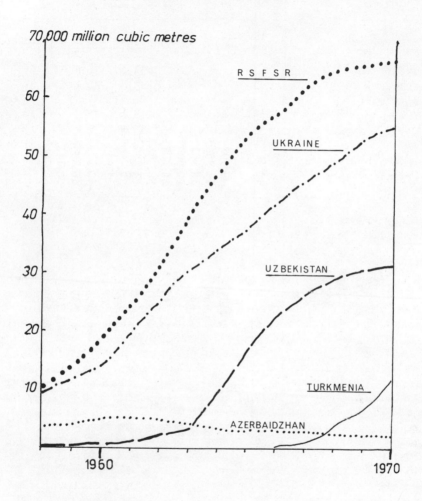

70,000 million cubic metres

Source: Table 2.13.

TABLE 2.13

Output of Natural Gas (without Oil-Well Gas),
by Republic, 1958-70
(in millions of cu.m.)

Year	USSR	RSFSR	Ukrainian SSR	Azerbaidzhan SSR
1958	22,658	10,144	9,102	3,331
1959	28,852	13,690	11,264	3,744
1960	37,594	18,834	13,801	4,541
1961	50,382	24,420	19,998	4,880
1962	63,509	31,487	25,307	4,616
1963	77,698	40,391	30,313	4,021
1964	94,355	47,528	34,233	3,054
1965	111,184	53,699	37,786	3,030
1966	125,173	57,333	41,913	3,068
1967	138,587	61,998	45,579	2,896
1968	149,520	64,640	48,985	2,486
1969	159,531	65,716	53,143	2,379
1970	170,000	66,800	55,000	2,200

Year	Uzbek SSR	Turkmen SSR	Kirghiz SSR	Tadzhik SSR	Kazakh SSR
1958	81	—	—	—	—
1959	150	4	—	—	—
1960	387	—	31	—	—
1961	947	2	135	—	—
1962	1,948	16	137	—	—
1963	2,895	64	114	—	—
1964	9,223	153	136	23	—
1965	16,367	103	147	52	—
1966	22,453	151	155	90	9
1967	26,524	1,077	250	245	17
1968	28,890	3,634	285	363	237
1969	30,669	6,298	335	433	558
1970	31,400	11,800	300	500	1,900

Sources: Lvov, M. S., Resursy prirodnogo gaza SSSR, Moscow 1969: 26-27; Gazovaya promyshlennost no. 4, April 1970: Appendix Table 3; RSFSR v tsifrakh v 1971 g, Moscow 1972: 27; Ekonomicheskaya gazeta no. 8, February 1973: 2.

TABLE 2.14

Important Areas of Natural Gas Production in RSFSR, 1958-70

(in percentage of total USSR output)

Area	1958	1965	1966	1967	1970
RSFSR	44.8	48.4	45.7	44.7	42.0
Krasnodar krai	2.9	20.0	20.0	19.0	15.0
Stavropol krai	18.7	13.6	12.5	11.5	8.0
Volgograd oblast	8.0	6.0	4.8	4.1	3.0
Saratov oblast	6.5	6.0	5.0	3.8	3.0
Tyumen oblast	—	—	0.5	3.7	5.0

Source: Lvov, M. S., Resursy prirodnogo gaza SSSR, Moscow 1969: 26-27; 1970 figures estimated by author.

rising to a possible 60,000 million, which will make it one of the most important gas fields in the USSR. In 1971 it produced about 3,000 million. The Komi ASSR is also a most promising area.

The amount of natural gas extracted in the Ukraine has risen steeply every year, but its contribution as a percentage of the total USSR production has dropped from 40 percent in 1958 to 33 percent in 1967, since when it has dropped below 30 percent. In the Ukraine 65,000 million cu.m. were extracted from about fifty gas and gas condensate fields in 1971, and in 1973 the output was expected to rise to 66,900 million. The most important of these fields has been the Shebelinskoe, where in 1971 alone some 31,000 million cu.m. were produced, making a total production over fifteen years of more than 330,000 million cu.m. By May 1972 the Efremovskoe field had supplied 20,000 million cu.m. in five years of operation.

Since 1964 the third major producing republic has been Uzbekistan, the contribution of which rose from a mere 4 percent in 1958 to 19 percent in 1967. The extraction of natural gas is no longer increasing at the same rate, and Uzbekistan's share in the total Soviet output is expected to fall below that of Turkmenia. Fourteen gas fields were exploited in the period from 1960 to 1971, the most important being the Gazlinskoe, which produces some 25,000 million cu.m. a year. In 1971 the Mubarek group of fields produced 4,000 million cu.m., the Uchkyrskoe 1,800 million, and the Shakhpakhty about 2,000 million.

Fourth in importance in 1972 was Turkmenia, the share of which rose from .9 percent in 1967 to over 7 percent in 1960. This republic had produced a total of 60,000 million cu.m. by February 1973. Recent discoveries here, especially the Shekhitlinskoe (Shatlyk) and Naipskoe fields, should allow eventual output to reach 75,000 to 100,000 million cu.m., with about 65,000 million being produced in 1975. Most of this (56,000 million) will be produced in the east of the republic by the "Turkmengazprom" association, with 9,100 million being produced in the west by the "Turkmenneft" association. Output in 1972 was 19,845 million cu.m. and was expected to be 28,600 million cu.m. in 1973.

The contribution of Azerbaidzhan has dropped considerably from 1958 (14.6 percent) to 1970 (under 2 percent,) being overtaken by that of Uzbekistan in 1964 and that of Turkmenia in 1968. Azerbaidzhan is also the only gas-producing republic where the actual amount extracted has been decreasing over the past decade.

The Kazakh, Tadzhik, and Kirgiz republics are steadily increasing their yearly output of natural gas, but as of 1973 the total was less than 3 percent of the extraction figure for the USSR as a whole. In 1973, when the Tenge field was joined to the central Asia-Center pipeline, Kazakhstan was expected to produce 5,000 million cu.m.[58]

The extraction of oil-well gas is shown in Table 2.15. Over two-thirds is produced in the RSFSR; the remainder is contributed by Azerbaidzhan, the Ukraine, and Turkmenia. The Kazakh, Uzbek, Tadzhik, Kirgiz, and Belorussian republics contribute very little to the total for the USSR as a whole, although the production of oil-well gas has some local significance. Recent increases in Belorussia are especially worth mentioning, since this area has long been without any local supply of the more valuable fuels.[59]

By far the greater part of liquid gas production is again in the RSFSR, with less important contributions from Azerbaidzhan and the Ukraine.[60] (See Table 2.16.)

The rapid expansion of the gas industry has been the result of greatly increased production of natural gas, while for over a decade there has been little change in the quantity of gas manufactured from coal and oil shales because of the far higher costs involved.[61] About two-thirds of the total is manufactured in the RSFSR and most of the remainder in Estonia. (See Table 2.17.)

Certain conclusions may be drawn from a comparison of the distribution of gas production in the USSR with the location of gas reserves given in the previous section.

First, the expansion of reserves in Saratov oblast, Krasnodar krai, and Stavropol krai is not keeping pace with extraction, and production is therefore likely to decrease in these areas in future years.

TABLE 2.15

Output of Oil-Well Gas 1950-70, by Republic
(in millions of cubic metres)

	1950	1960	1962	1965	1966	1967	1968	1969	1970 (Estimated)
USSR	1,799	7,706	10,014	16,483	17,789	18,858	19,581	21,570	22,700
RSFSR	589	5,579	6,785	10,559	11,709	12,783	13,707	15,275	16,545
Ukrainian SSR	87	484	851	1,576	1,704	1,863	1,957	2,259	2,100
Azerbaidzhan SSR	1,000	1,300	1,990	3,149	3,105	2,875	2,507	2,559	2,300
Uzbek SSR	51	59	86	108	113	114	98	97	90
Turkmen SSR	65	234	238	1,055	1,114	1,149	1,209	1,224	1,375
Kirghiz SSR	–	10	18	8	8	6	6	6	6
Tadzhik SSR	–	–	–	–	–	–	3	5	9
Kazakh SSR	7	39	46	29	37	66	84	122	100
Belorussian SSR	–	–	–	–	–	2	10	24	175

Source: "Pokazateli razvitiya gazovoi promyshlennosti SSSR," Table 4, Gazovaya promyshlennost no. 4, April 1970.

TABLE 2.16

Liquid Gas Production in 1969, by Republic
(in tons)

Area	Production
USSR	4,208,300
RSFSR	3,841,800
Azerbaidzhan SSR	136,100
Ukrainian SSR	133,800
Turkmen SSR	39,400
Belorussian SSR	23,400
Uzbek SSR	14,500
Kazakh SSR	12,800
Georgian SSR	6,500

Source: Gazovaya promyshlennost no. 4, April 1970: Table 6.

TABLE 2.17

Manufacture of Gas from Coal and Oil Shale in USSR
(in millions of cu.m.)

Year	USSR	RSFSR	Estonian SSR
1940	173	167	1.7
1950	420	226	173
1960	1,911	1,488	433
1965	1,690	1,175	515
1966	1,713	1,174	539
1967	1,735	1,170	565
1968	1,713	1,156	557
1968	1,719	1,138	581
1970	1,634	1,053	581

Sources: Narodnoe khozyaistvo SSSR v 1970 godu, Moscow 1971.
"Pokazateli razvitiya gazovoi promyshlennosti SSSR," Table 5, in
Gazovaya promyshlennost no. 4, April 1970.

Second, the expansion of reserves in all other gas-bearing areas has been sufficient to permit a considerable increase in the rate of extraction.

Third, in 1970 the ratio of output to reserves in Komi ASSR, Orenburg oblast, Sakhalin oblast, Siberia, Kazakh SSR, Turkmen SSR, and Kirgiz SSR was extremely low compared with other areas. It is therefore to be expected that output will rise more quickly in the above-mentioned areas. Because of their particularly large reserves, Tyumen oblast and Turkmen SSR should soon be producing a large proportion of the total USSR output.

Finally, because these reserves, which have barely been tapped, lie preponderantly in the less populated eastern part of the Soviet Union, it must be decided whether to utilize them by building appropriate industrial centers near the sources of energy or to transmit the gas to established areas of demand by extending the system of gas pipelines described below.

The Directives of the 24th Congress of the CPSU on the development of the economy up to 1975 support these conclusions. Gas output is expected to reach 300,000 to 320,000 million cu.m. in 1975, with greater efficiency in extracting natural gas, in utilizing oil-well gas, and in increasing production of liquid gases. Over 30,000 kilometers of gas pipeline are to be constructed to ensure the supply of gas from the new extraction areas to the European parts of the USSR. Output in the RSFSR should reach 141,000 million cu.m. by 1975. A new center for the extraction and processing of natural gas is to be created in Orenburg oblast. The exploitation of gas fields in the north of Tyumen oblast is to be accelerated, and processing plants with a yearly capacity of up to 6,000 million cu.m. are to be constructed in West Siberia. In the Ukraine "the exploration and exploitation of new gas fields is to be accelerated," and in Uzbekistan there is to be a "significant development" of the gas industry; neither expression, however, is quite so concrete as the claim made for Turkmenia, where the yearly extraction of natural gas is to increase 4.3 to 4.7 times over the five years of the plan.[62]

It is planned to raise production by concentrating on the largest fields, which in 1975 should yield the following amounts of natural gas: Naipskoe, 25,000 million cu.m.; Shatlykskoe, 32,000 million; Orenburg, 30,000 million; Medvezhe, 35,000 million; and Vuktylskoe, 15,000 million.[63] Present growth in gas production by republic is shown in Table 2.18.

In 1973, however, it did not seem certain that output could reach 320,000 million cu.m. by 1975, since production in 1972 and 1973 was considerably below the original plan. Nonetheless, a steady growth in the extraction of natural gas is being achieved.

TABLE 2.18

Production of Natural Gas (Including Oil-Well Gas),
by Republic, 1970-75

Area	1970 (actual)	1971 (actual)	1972 (actual)	1975 (planned)
USSR	198.0	212.0	221.4	320.0
RSFSR	83.0	87.4	87.4	144.6
Ukrainian SSR	60.0	64.7	67.2	62.0
Turkmen SSR	13.1	16.9	21.3	65.1
Uzbek SSR	32.1	33.6	33.7	33.7
Kazakh SSR	2.1	2.7	3.5	7.0
Azerbaidzhan SSR	5.5	5.8	6.9	6.0
Tadzhik SSR	0.4	0.4	0.5	0.5
Belorussian SSR	0.2	0.3	0.4	0.6
Kirgiz SSR	0.0	0.4	0.4	0.6

NA: Not available.

Sources: Gazovaya promyshlennost no. 12, December 1972: 6;
SSSR i soyuznye respubliki v 1971 godu, Moscow 1972; SSSR i soyuznye
respubliki v 1972 godu, Moscow 1973.

Transportation and Storage

The System of Pipelines

Any increase in the output of natural gas must be directly proportional to progress in methods of transmission, storage, and utilization. It is therefore to be expected that the rapidly expanding output
figures discussed above should be matched by those for the construction of pipelines, given in Table 2.19. Again, Figure 2.5 clearly shows
the accelerated construction that began in the late 1950s.

Nonetheless the USSR still lags behind the United States in
this sphere, as may be seen from the comparative figures in Table
2.20. Again, there seems to be no reason to expect that the United
States will lose its lead in the near future, since yearly construction
has often been three times that of the USSR.[64] (See Table 2.21.)

Certain factors must be borne in mind when making these comparisons. The USSR has a much higher percentage of major gas

TABLE 2.19

Cumulative Growth of Major Pipelines in USSR, 1950-75

Year (January)	Length (in km.)	Throughput (in millions of cu.m.)	Year (January)	Length (in km.)	Throughput (in millions of cu.m.)
1950	2,273	1,500	1966	42,273	—
1955	4,279	—	1967	47,550	—
1956	4,861	—	1968	52,748	155,100
1958	9,633	—	1969	56,088	166,000
1959	12,202	—	1970	60,334	181,500
1960	16,494	32,800	1971	67,500	200,300
1961	20,983	—	1972	72,500	223,500
1962	25,328	—	1973	78,500	—
1963	28,492	—	1974*	86,000	—
1964	33,033	—	1975*	100,000	
1965	36,908	—			

*Planned

Sources: "Pokazateli razvitiya gazovoi promyshlennosti SSSR," Table 11, Gazovaya promyshlennost no. 4, April 1970; Stroitelstvo truboprovodov no. 1, January 1971: 2; Gazovaya promyshlennost no. 3, March 1972: 46; Narodnoe khozyaistvo SSSR 1922-1972, Moscow 1972: 306.

TABLE 2.20

Cumulative Growth of Major Pipelines in USSR
and United States
(length in km.)

Year	USSR	United States
1950	2,273	176,000
1955	4,279	229,300
1960	16,494	292,500
1965	36,908	360,000
1970	60,334	428,000
1975 (planned)	100,000	465,600

Sources: Yufin, V. A., et. al., Truboprovodnyi transport 1968-1969, Moscow 1970: 28; Statistical Abstract of the United States, 1971, Washington 1971: 504.

FIGURE 2.5

Cumulative Construction of Major Gas Pipeline Network
in USSR, 1950–1974

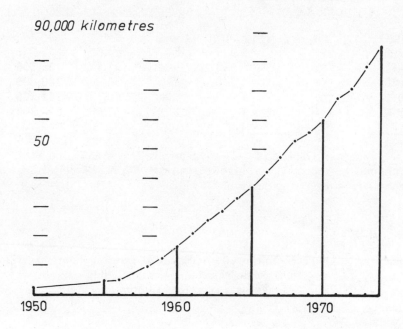

Source: Table 2.19.

TABLE 2.21

Annual Construction of Major Gas Pipelines in USSR and
United States, 1965-70
(length in km.)

Year	USSR	United States
1965	5,365	10,197
1966	5,277	13,755
1967	5,198	15,059
1968	3,340	12,896
1969	4,246	12,110
1970	7,000	11,123

Sources: Grokhotova, T. V., Shor, L. D., "Analiz izmenenii stoimosti truboprovodnogo stroitelstva v S. Sh. A.," Stroitelstvo truboprovodov no. 1, January 1971: 36.

TABLE 2.22

Increase in the Use of Large Diameter Pipe
in the Gas Pipeline Grid, 1962-69
(as percentage of whole)

Diameter of Pipeline (in mm.)	1962	1966	1969
1,220	—	—	1
1,020	5	18	24
820	12	10	9
720	29	25	21
529	20	20	19
426	5	5	4
377	3	3	3
325	13	10	9
259-273	5	3	4
219	4	3	4

Source: "Pokazateli razvitiya gazovoi promyshlennosti SSSR," Table 8, in Gazovaya promyshlennost, no. 4, April 1970.

pipelines of very large diameter pipe than the United States does. In 1970, for example, under 30 percent of the transmission pipelines in the United States were above 500 mm. in diameter, as compared with over 70 percent in the USSR.[65] In the Soviet Union the mileage of the distribution network is less than that of the transmission network (55,787 km. to 60,334 km. in 1970).[66] This can be explained by the fact that most of the gas extracted in the USSR is dispatched to large consumers such as power stations and industrial plants, while a relatively small proportion is allocated to domestic consumption.

Since the extensive building of pipelines was so long delayed in the USSR, most of the construction undertaken has had the advantage of the latest technology. This accounts to some extent for the preponderance of large diameter pipe in the grid. Even in 1962 over 5 percent of the gas pipeline system was constructed of 1,020 mm. diameter pipe, and by 1969 this had risen to 24 percent. (See Table 2.22.) The grid has expanded in recent years, using mostly the larger diameter pipelines.[67]

By increasing the diameter of the pipeline, capital investment and transport costs can be lowered. One pipeline of 2,520 mm., with a throughput of 100,000 million cu.m. a year, can replace ten earlier pipelines of 1,020 mm. in diameter, while using only half as much metal. Progress in pipeline technology is being accompanied by new methods of extracting gas from high-productivity wells of increased diameter. Wells being bored in Tyumen oblast are 200 to 250 mm. in diameter instead of the present average of 100 to 155 mm. Such wells could yield up to 5 million cubic meters daily, compared with the present level of 700,000 cubic meters. Boring 150 high-productivity wells in the Urengoiskoe field rather than 600 ordinary wells should make it possible to save over 50 million rubles.[68]

The network of major pipelines is shown on Map 2.2. It consists of the following systems of pipelines.

Central System. Gas from the Stavropol and Krasnodar fields is transmitted to the industries of the Center and via Rostov, Taganrog, and Amvrosievka to Donetsk. From the Center it continues to the Northwest by the Serpukhov-Leningrad pipeline. Its yearly capacity is over 40,000 million cu.m.

East Ukrainian System. From the fields at Shebelinka, gas pipelines stretch via Kharkov, Belgorod, and Bryansk to Moscow, via Poltava to Kiev, and via Dnepropetrovsk and Krivoi Rog to Odessa. The Shebelinka-Ostrogozhsk pipeline, which has a yearly capacity of over 26,000 million cu.m., connects this system with the Central System.

MAP 2.2

Major Natural Gas Pipelines

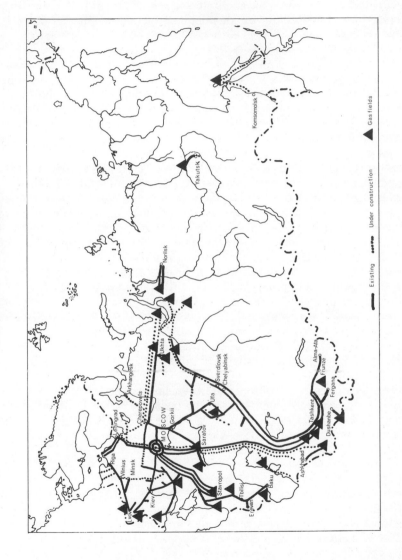

Source: Sovetskii Soyuz—Obshchii obzor, Moscow 1972.

Western System. Gas is supplied from the West Ukrainian fields to Belorussia, Lithuania, and Latvia through the Dashkova-Kiev and the Dashkova-Minsk-Riga pipelines, which can transport over 6,000 million cu.m. yearly.

Volga System. Gas from the Volgograd and Saratov fields is transmitted through the following pipelines: Saratov-Moscow, Saratov-Gorkii-Cherepovets, Saratov-Volsk, Korobki-Storozhevka, and Minnibaevo-Kazan-Gorkii. The Volga System has a yearly capacity of more than 6,000 million cu.m.

Caucasus System. Gas from the Stavropol, Krasnodar and Azerbaidzhan fields is transported by the following pipelines: Stavropol-Nevinnomysk-Groznyi, Kradag-Akstafa-Tbilisi-Erevan, Maikop-Nevinnomysk, and Ordzhonikidze-Tbilisi. Their yearly capacity is over 5,000 million cu.m.

Central Asian System. Gas from Dzharkak is transported to Bukhara, Samarkand, Tashkent, Frunze, and Alma Ata. By 1975 its yearly capacity is expected to reach 22,000 million cu.m.

Ural System. The pipeline from the Gazli fields to the industries of the Urals is over 2,000 km. in length and supplies gas to the towns of Chelyabinsk, Sverdlovsk, Nizhnii Tagil, Magnitogorsk, Novotroitsk, and others. It is joined by the Igrim-Serov pipeline at Nizhnii Tagil. Yearly capacity is around 30,000 million cu.m.[69]

Central Asia-Center System. The double pipeline from Central Asia to the Center is over 2,750 km. in length and has a yearly capacity of about 25,000 million cu.m. Pipes of 1,220 mm. diameter were used in building the second line. Two more lines are planned, to increase capacity by 65,000 million cu.m. in 1975. The fourth line will be from the new Shekhitlinskoe field and will use pipe of 1,420 mm. diameter as far as Aleksandrov Gai. Five compressor stations with a total capacity of 30 MW were being built in 1973 to help maintain pressure at 55 atmospheres.

"Siyanie Severa" ("Northern Lights") System. This network, consisting of over 4,500 km. of pipeline, is planned to transmit gas from the West Siberian fields to the central and western areas of the USSR. The stretch from the Vuktylskoe field in Komi ASSR to Torzhok has been constructed of 1,220 mm. diameter pipe. The second line from Ukhta to Torzhok has also been completed. The 1,220 mm. diameter pipeline has a yearly capacity of about 15,000 million cu.m., some 50 percent greater than pipeline of 1,020 mm. It is also planned to use pipeline of up to 2,520 mm. diameter, since it has been calculated that by using a double pipeline of 2,020 and 2,520 mm. diameter, with a yearly capacity of up to 130,000 million cu.m., it will be possible to save about 3 million tons of metal, compared with the equivalent fifteen lines of 1,020 mm. pipeline. Another pipeline is being built over the permafrost areas from Nadym to Punga and on to the Center. By 1975 this pipeline system (Medvezhe-Nadym-Punga-Nizhnyaya Tura-Perm-Gorkii-Center) is to have an annual capacity of 38,000 million cu.m. There were reports in April 1972 that delays were being caused by hasty, inefficient welding and failure to observe rules for laying pipes in marshy ground, but that progress continued to be made.[70]

In building the Nadym-Ukhta-Torzhok section it was decided that two pipelines of 1,420 mm. diameter and two of 2,520 mm. diameter would be used. Between Ukhta and Gryazovetsk there will be two pipelines of 1,220 mm. diameter, one 1,420 mm. long and two 2,520 mm. long. The section from Gryazovetsk to Torzhok will consist of several lines of 1,220 mm. and 1,420 mm. diameter, and the section from Nadym to Punga of 1,220 mm. and 2,020 mm. diameter pipe. Work has proceeded more slowly than planned. On 18 February 1973 Pravda revealed that in 1972 only 170 kilometers out of a planned 416 had been laid and that two compressor stations out of a planned four had been built.[71]

In 1969 the most northern pipeline in the world was constructed from Messoyakha to Norilsk, calling for the solution of many complex technical and logistic problems, such as in welding large-diameter pipes where temperatures can drop to -50° C and where the only transport even for heavy pipes is often by aircraft.

In March 1973 a pipeline of 186 kilometers was completed, joining the Sredne-Vilyuiskoe field in Eastern Siberia to the Taas-Tumus-Yakutsk pipeline. Welding continued even at -50°C. Before 1980 it is planned to extend this pipeline over 2,100 km. from Yakutsk to Magadan; it will have an annual capacity of 30,000 million cu.m.[72]

Producing large-diameter pipe calls for a very high level of technology, and at first most of the pipe used was imported. Nevertheless, Soviet technologists have made great progress in recent years, and the first pipes of 1,220 mm. diameter were produced in a Chelyabinsk plant in 1967. By 1970 the Novomoskovskii metallurgy plant had mastered the production of 2,520 mm. diameter pipe. Laying and welding pipelines of such size has necessitated building a new type of machinery.[73] Compressor capacity has been very low, and throughout has in general been well below the possible level.

Some of the first pipelines functioned by layer pressure alone. As this fell, an increasing number of compressor stations had to be built, and delays in their construction frequently resulted in significant drops in the amount of gas delivered to consumers. A fall of one atmosphere in layer pressure can result in the loss of a million cubic meters delivered to consumers per day. At the huge Gazliiskoe field in Uzbekistan, for example, pressure has dropped from 55 to 40 atmospheres and is continuing to fall. Yet an additional compressor station planned for completion in 1970 had still not been finished by spring 1973. To increase the efficiency of the pipelines, additional compressor stations with gas turbines of 6,000 and 10,000 KW have recently been installed. Pipelines of 1,420 mm. and 2,520 mm. diameter will have turbines of 16,000 and 25,000 KW.[74]

Recent additions to the gas pipeline network include the following lines: Maiskoe-Ashkhabad-Bezmein, Maiskoe-Bairam Ali (to be completed in 1974), Efremovka-Dikanka-Kiev-West Ukraine, Kuleshovka-Melekess-Ulyanovsk, Mokrous-Tolyatti, Ostrogozhsk-Beloussovo, Valdai-Pskov-Riga, and Messoyakha-Norilsk (second line). In 1972 work began on an 810 km. pipeline from Shebelinka to Dolina (on the Czechoslovak border) through Dikanka and Kiev, using 1,420 mm. diameter pipe.[75]

In the three years following the opening of the "Bratstvo" ("Fraternity") pipeline in August 1967, over 2,000 million cu.m. of natural gas were transmitted to Czechoslovakia. Its basic diameter is 820 mm., and its yearly capacity is now 4,500 million cu.m. A second line was constructed of 1,420 mm. diameter pipe, working at a pressure of 75 atmospheres. Planning began in summer 1973 for a third line, branching off at Svalyava and running through Leninvaros to Budapest.

About 3,000 km. of pipeline are being laid to extend supplies to Hungary, Poland, East Germany, Bulgaria, and Western Europe. By 1975 Comecon nations will be receiving up to 15,000 million cu.m. annually.[76]

Import and Export

In December 1969 a twenty-year contract was signed with the Italian state-owned oil and gas corporation, E.N.I. (Ente Nazionale Idrocarburi), by which the USSR is to supply Italy with 110,000 million cu.m. of natural gas at between 1,200 to 6,000 million cu.m. per year starting in 1973. The Italians have built a pipeline across Austria to meet "Bratstvo" at the Austrian-Czech border. The bulk of the goods that the USSR will receive in exchange will be 1,220 mm. diameter steel pipes to supplement those produced in Soviet factories.

Negotiations have also been going on with Austria, West Germany, and France. The flow of gas to Austria, which will be supplied with 1,500 million cu.m. of Soviet gas, began in 1971. In February 1970 a contract was signed with the German company "Ruhrgas" to supply West Germany with 52,000 million cu.m. over a period of twenty years, from 500 million cu.m. in the first year up to 3,000 million cu.m. per annum. In September 1969 an agreement was made with the state-owned "Gaz de France" for the delivery of up to 3,000 million cu.m. of natural gas to the French consumer.

In April 1970 Finland signed a twenty-year contract, to become operative in 1974, by which the USSR will supply initially 500 million cu.m. of natural gas a year, rising to 3,000 million cu.m. by the end of the period. The pipeline to Kotka was constructed by Soviet experts.

In 1972 Japan started negotiations with the USSR to discuss possible cooperation in exploiting the rich natural gas deposits in Sakhalin Island and its continental shelf. The Japanese have also expressed interest in Siberian gas.

In view of the warmer relations between the Soviet Union and the United States since the summer of 1972, it is now very likely, if political obstacles can be overcome, that liquefied natural gas from Siberia will be shipped to America to help solve the problem arising there from exhaustion of reserves. This would help to pay for imports of industrial plant equipment from the United States.[77]

The credit granted to the USSR in return by Austria, Italy, West Germany, and France will be used to purchase some of the advanced equipment and material of which there is a shortage in the Soviet Union. This will include telecommunication and TV equipment, compressors, valves, and road-building machinery. Although the USSR produced nearly 12 million tons of steel tubes in 1969, this was not enough to meet construction plans for the pipeline system, and the resulting discrepancy is being made up by importing steel tubes. In 1970, for example, the USSR imported 1,338,600 metric tons of steel tubes and pipes. Finland is to supply 60,000 to 70,000 tons over a ten-year period, starting from 1971.[78]

Trade of a reverse order takes place between the USSR and two of the underdeveloped countries of the southern border, Iran and Afghanistan.

The Soviet contribution accounts for about 60 percent of foreign aid to Afghanistan and includes the construction of a thermoelectric power plant of 36,000 KW capacity and a chemical fertilizer factory at Mazar-i-Sharif. Natural gas deposits discovered about 15 km. east of the town of Shibargan have been exploited with Soviet aid, and Afghanistan will repay most of her debt through the export of natural gas to the USSR over fifteen years. Soviet engineers have built a pipeline of 820 mm. diameter some 98 km. from Shibargan to the Soviet border and across the Amu-Darya river, and have also constructed a pipeline of 325 mm. diameter over 88 km. to the chemical fertilizer plant at Mazar-i-Sharif. By 1973 over 11,000 million

cu.m. had already been imported by the USSR, the amount received
having increased each year as follows:[79]

Year	Amount Imported (in millions of cu.m.)
1967	207
1968	1,500
1969	2,030
1970	2,500

The Trans-Iranian pipeline is to supply up to 10,000 million
cu.m. yearly to the Caucasus. Total deliveries by February 1973
amounted to 13,000 million cu.m. Soviet engineers constructed the
487 km. section from Sava to Astara of 1,016 mm. diameter pipe and
the 112 km. section to Teheran of 762 mm. diameter pipe.[80]

Although there are obvious political advantages to be gained
in expanding Soviet influence south to the Indian Ocean, it can be
claimed that these transactions make perfectly good economic sense
and do not necessarily conceal any motive other than a desire to
increase foreign trade. Since pipelines to Eastern Europe are already
in existence, and since the USSR must purchase certain goods and
equipment in the West, there is a sound economic argument for buying
gas cheaply from Iran and Afghanistan and then selling gas at a higher
price in Western Europe.

Seasonal Fluctuations in Demand

Because of daily and seasonal fluctuations in demand, planners
in the gas industry must calculate the optimum balance between transmission, storage, and the alternative off-peak utilization of gas to replace other fuels. If the pipeline system is capable of meeting the full
demand for gas at periods of maximum consumption it will certainly
only be partially utilized at off-peak times.

An obvious solution to this problem is to increase the available
storage capacity in the main areas of consumption. Normal surface
storage methods necessitate enormous capital investment and consume quantities of valuable metal. The use of underground storage
is therefore spreading in the Soviet Union. The cheapest method is
to use exhausted natural gas reservoirs, such as those in the Kuibyshev and Saratov fields (Bashkatovskoe, Amanakskoe, and others).
Unfortunately, in the Soviet Union most of these natural reservoirs
are situated too far from the areas of consumption to be useful, and
Soviet specialists are therefore making artificial reservoirs in suitable geological formations, such as water-bearing layers in which

the water can be displaced by gas. For the normal functioning of this type of storage it is first necessary to pump an equal amount of gas that will not be recovered into the reservoir, to act as a "cushion" for the recoverable stored gas. Such storage reservoirs are already in use in the Moscow, Leningrad, and Kiev areas, and others are being formed near Riga, Sverdlovsk, Novgorod, and Tashkent.[81]

The earliest was the Kaluga reservoir, into which gas was first pumped in the autumn of 1959. It was brought into commercial exploitation in 1963 with a total capacity of 400 million cu.m., with active storage for 200 million. It has thirteen wells with a maximum delivery rate of 9.5 million cu.m. daily.

The Shchelkovo reservoir, situated 17 km. northeast of Moscow, first received gas in 1961. It has a total capacity of 2,800 million cu.m., with active storage for 1,500 million. In 1968 its 50 wells gave a maximum delivery rate of 14 million cu.m. daily.

The Olishevko reservoir, some 100 km. to the north of Kiev, is not yet in full exploitation, but in the extremely cold winter of 1968-69 over 114 million cu.m. were removed for consumption at a maximum daily rate of 1.1 million cu.m. Its active capacity should reach 250 million cu.m.

The Poltoratsk reservoir went into experimental exploitation in August 1965 and supplied over 96 million cu.m. to the consumer in the winter of 1968-69. Its active capacity should reach 380 million cu.m.

The Kanchura reservoir in the Bashkir ASSR will help to eliminate seasonal fluctuations in the loading of the Bukhara-Urals pipeline.

The Kolpino reservoir in the Leningrad area supplied 45.7 million cu.m. over the winter of 1968-69 and is expected to reach an active capacity of about 100 million cu.m.

The Gatchino reservoir, which is also in the Leningrad area, was able to deliver 184 million cu.m. at peak periods of consumption in its sixth yearly cycle of exploitation (1968-69).

By 1972 there were three underground stores in Kuibyshev oblast, with a total capacity of 200 million cu.m. The largest reservoir is in Kuibyshev oblast, opened in summer 1973. It can take, or supply, up to 30 million cu.m. a day and will have a total capacity of 4,000 million cu.m.[82]

In 1971 about 6,000 million cu.m. were pumped into underground storage. Using these reservoirs has greatly improved winter supplies of gas to industry and homes, especially during sudden cold spells. It has been calculated that the cost of forming the Shchelkovo reservoir was only a third that of building a new pipeline capable of bringing an equal supply of gas from the nearest field. Moreover, the yearly running costs of a pipeline would be 1.8 times those of the reservoir. It is planned to expand the Shchelkovo and other existing reservoirs, and also bring into exploitation new underground storage reservoirs near Baku, Groznyi, Volgograd, and other areas of high consumption.[83]

Another solution to the problem of seasonal fluctuation in demand is to burn surplus gas in electric power stations in summer and thus utilize the pipeline system more fully. In such cases gas is alternated with other fuels that can be more easily stored on the site.[84]

Recent Innovations

Various new policies and technical innovations are being introduced to improve the efficiency of the gas supply.

The separate gas pipeline systems are being joined into a single national grid with a centralized automated control. The first stage is to have a single system for the European USSR, to which the Central Asian and Urals pipelines will eventually be linked. Several hundred gas fields have already been connected by over 70,000 km. of pipeline to some 1,570 towns. Further computerization and automation of the national grid should simplify the rational integration of the gas industry into the energy economy as a whole.[85]

New ways of constructing pipelines under difficult conditions are being initiated after testing in the scientific research institutes of the gas ministry. Underwater construction methods have been improved, allowing the Kuleshovka-Melekess-Ulyanovsk pipeline to cross the 6 km. of the Kuibyshev water reservoir. In polar winter conditions 4 km. were laid across the Enisei river at depths up to 47 m. for the Messoyakha-Norilsk pipeline. Machines capable of cutting ice up to 2 m. thick are now being used.[86]

In the trackless wastes of the Far North and in the deserts of Central Asia it has been necessary to supply pipeline construction sites with pipes and other equipment by helicopter and airplane. Strong arguments are being put forward for the reintroduction of transport dirigible airships, which would have much greater range and cargo capacity than either helicopters or airplanes.[87]

Further research is being carried out to examine the practical possibilities of transporting gas at low temperatures (as low as -70° C) or even in a liquid state at -110°C. While this would increase the throughput, it is not yet clear that it would be commercially viable.[88]

It is to be expected that the length of pipe in the gas pipeline grid and the numbers of underground reservoirs will continue to grow in order to increase the flow of fuel from the more remote fields to the industrial and population centers of the USSR and to make the supplying of gas more flexible and reliable. The utilization of natural gas in the Soviet economy is discussed in Chapter 8.

NOTES

1. Hodgkins, J. A., Soviet Power: Energy Resources, Production and Potential, London 1961: 135.

2. Kortunov, A. K., in Sotsialisticheskaya industriya, 5 September 1970: 1.

3. Stroitelstvo truboprovodov no. 1, January 1971: 1.

4. Lvov, M. S., Resursy prirodnogo gaza SSR (Natural gas resources of the USSR), Moscow 1969: 11; Melnikov, N. V., Mineralnoe toplivo (Mineral fuel), Moscow 1971: 93.

5. Gazovaya promyshlennost no. 3, March 1972: 46; See Table 2.10.

6. Lvov, op. cit.: 11.

7. Ibid.: 13.

8. Ibid.: 15.

9. Kortunov, A. K., et. al., Gazovaya promyshlennost SSSR (Gas industry of the USSR), Moscow 1967: 3; Gazovaya promyshlennost no. 12, December 1972: 5.

10. Bokserman, Yu. I., "Razvitie gazovoi promyshlennosti SSSR" (Development of the USSR gas industry), Neftyanik no. 2, February 1970: 6.

11. Kortunov, 1967, op. cit.: 4.

12. Neftyanik no. 2, February 1970: 6.

13. Kortunov, 1967, op. cit.: 4.

14. Lvov, op. cit.: 17.

15. Kostrin, K. V., "70 let otechestvennomu trudoprovodostroeniyu" (70 years of Soviet pipeline construction), Stroitelstvo truboprovodov no. 1, January 1970: 7.

16. Kortunov, 1967, op. cit.: 5-6.

17. Narodnoe khozyaistvo SSSR v 1968 godu (National economy of the USSR in 1968), Moscow 1969: 233.

18. Kortunov, 1967, op. cit.: 6-7.

19. Lvov, op. cit.: 20.

20. Energeticheskie resursy SSSR: Toplivno-energeticheskie resursy (Energy resources of the USSR: fuel resources), Moscow 1968: 453.

21. Kortunov, 1967, op. cit.: 1-7.

22. Lvov, op. cit.: 23.

23. Kortunov, 1967, op. cit.: 8.

24. Energeticheskie resursy, op. cit.: 457.

25. Kortunov, 1967, op. cit.: 8.

26. References to this classification of reserves may be found in most Soviet works on fuel resources. See also, in English, R. W. Campbell, The Economics of Soviet Oil and Gas, Baltimore 1968: 60-62, 198-99.

27. Dzhalaev, M. Ts., "Rol prirodnogo gaza v razvitii mirovoi energetiki" (The role of natural gas in the development of world energy production), Gazovaya promyshlennost no. 1, January 1972: 47-51; Pominov, V. F., "Mirovye vyyavlennye zapasy prirodnogo i neftyanogo poputnogo gazov" (World reserves of natural and incidental gases), Gazovaya promyshlennost no. 1, January 1971: 48-49.

28. Energeticheskie resursy, op. cit.: 34, 348, 412.

29. Vasilev, V. G., "Syrevaya baza gazovoi promyshlennosti i perspektivy ee rasshireniya" (The raw material basis of the gas industry and prospects for its expansion), Gazovaya promyshlennost no. 4, April 1970: 10.

30. Ibid.: 8.

31. Ibid.: 9.

32. Gazovaya promyshlennost no. 1, January 1971: 1.

33. Galaktionov, B. V., "Prirodno-klimaticheskie i geokriologicheskie usloviya osvoeniya gazovykh mestorozhdenii severa zapadnoi Sibiri" (Natural climatic and geological conditions for the exploitation of gasfields in North West Siberia), Gazovaya promyshlennost no. 1, January 1970: 7-9.

34. Gazovaya promyshlennost no. 4, April 1970: 10; Pravda, 17 March 1972: 3; Pravda, 22 August 1973: 2.

35. Gazovaya promyshlennost no. 4, April 1970: 11.

36. Energeticheskie resursy, op. cit.: 350-67.

37. Gazovaya promyshlennost no. 4, April 1970: 10-12.

38. Energeticheskie resursy, op. cit.: 367-81.

39. Ibid.: 395-99.

40. Vasilev, V. G., "Syrevaya baza gazovoi promyshlennosti i perspektivy ee rasshireniya" (The raw material basis of the gas industry and prospects for its expansion), Gazovaya promyshlennost no. 4, April 1970: 12; Izvestiya, 20 March 1973: 1; Pravda, 22 August 1973: 2.

41. Pravda, 15 November 1972: 3; Energeticheskie resursy, op. cit.: 388-90.

42. Ibid.: 390-95.

43. Ibid.: 407-409; Stroitelstvo truboprovodov no. 12, December 1972: 18-24.

44. Gazovaya promyshlennost no. 4, April 1970: 10-12.

45. Trofimuk, A. A., "Razvitie gazovoi promyshlennosti Sibiri i Dalnego Vostoka" (Development of the gas industry in Siberia and the Far East), Gazovaya promyshlennost no. 4, April 1970: 14-15.

46. Gazovaya promyshlennost no. 8, August 1969.

47. Vasilev, V. G., "Osnovnye rezervy uvelicheniya prognoznykh zapasov gaza v SSSR" (Main reserves for increasing predicted resources of gas in the USSR), Gazovaya promyshlennost no. 1, January 1970: 3-4; Sovetskaya Rossiya, 31 March 1973: 1.

48. Ibid.: 5.

49. Ibid.: 6.

50. Ibid.: 7.

51. Brents, A. D., "Razvitiye gazovoi promyshlennosti i top-livnyi balans SSSR" (Development of the gas industry and the fuel balance of the USSR), Gazovaya promyshlennost no. 1, January 1969: 4.

52. Aleksandrov, A. V., "Osnovnye tendentsii nauchno-tekhnicheskogo progressa v gazovoi promyshlennosti" (Main trends of scientific and technological progress in the gas industry), Gazovaya promyshlennost no. 4, April 1970: 25-26.

53. Kortunov, A. K., "Gazovaya promyshlennost Sovetskogo Soyuza" (Gas industry of the Soviet Union), Gazovaya promyshlennost no. 8, August 1970: 2.

54. Gazovaya promyshlennost no. 1, January 1970: 1; Tolka-chev, A. S., Osnovnye napravleniya nauchno-tekhnicheskogo pro-gressa (The main trends in scientific and technical progress), Moscow 1971: 115.

55. Bokserman, Yu. I., "Razvitie gazovoi promyshlennosti SSSR" (Development of the gas industry of the USSR), Neftyanik no. 2, February 1970: 8.

56. United Nations Statistical Year Book 1970, New York 1971: 210-11; SSSR v tsifrakh v 1971 godu (USSR in figures in 1970), Moscow 1972: 60-61; Time, 12 June 1972: 35; Ekonomicheskaya gazeta no. 8, February 1973: 2.

57. Pravda, 19 October 1961: 3; Gazovaya promyshlennost no. 1, January 1966: 3; Ekonomicheskaya gazeta no. 8, February 1973: 1.

58. Gazovaya promyshlennost no. 3. March 1971: 2; Trofimich, A., et al., "Kak vzyat Tyumenskii gaz" (Exploiting Tyumen gas), Pravda, 28 March 1971: 3; Bataev, V., Klimenko, A., "Esli dobycha padaet" (If output falls), Sotsialisticheskaya industriya, 2 February 1971: 1; Vasilev, V., "Orenburgskiye sokrovishche" (Orenburg treasure-house), Pravda, 17 March 1972: 3; Denisovich, V., "Turk-menskiye Samotlory," Pravda, 6 May 1971: 3; Gazovaya promyshlen-nost no. 12, December 1972: 11-18; Sotsialisticheskaya industriya, 3 February 1973: 1; ibid., 13 January 1973: 2; Izvestiya, 23 March 1973: 3.

59. "Pokazateli razvitiya gazovoi promyshlennosti SSR," Table 4, in Gazovaya promyshlennost no. 4, April 1970.

60. Ibid.: Table 6.

61. Narodnoe khozyaistvo SSSR v 1968 godu, Moscow 1969; "Pokazateli razvitiya gazovoi promyshlennosti SSSR," Table 5, Gazo-vaya promyshlennost no. 4, April 1970.

62. Direktivy XXIV sezda KPSS po pyatiletnemu plany razvitiya narodnogo khozyaistva SSSR na 1971-1975 gody (Directives of the 24th Congress of the CPSU on the Five Year Plan for developing the national economy of the USSR, 1971-75), Moscow 1971: 18-19, 54-69.

63. Kortunov, A. K., "Zadachi gazovoi promyshlennosti v 1972 godu" (Problems of the gas industry in 1972), Gazovaya promyshlennost no. 2, February 1972: 3.

64. Grokhotova, T. V., Shor, L. D., "Analiz izmenenii stoimosti truboprovodnogo stroitelstva v S. Sh. A." (Analysis of changes in the cost of pipeline construction in the United States), Stroitelstvo truboprovodov no. 1, January 1971: 36.

65. Ibid.: 37; "Pokazateli razvitiya gazovoi promyshlennosti SSSR," Table 8, in Gazovaya promyshlennost no. 4, April 1970.

66. Radchik, I. I., Alekseev, G. A., "Kompleksnaya klassifikatsiya selskikh raionov" (Complex classification of country districts), Stroitelstvo truboprovodov no. 1, January 1971: 8; Statistical Abstract of the United States, 1971, Washington 1971: 504.

67. "Pokazateli razvitiya gazovoi promyshlennosti SSSR." Table 8, in Gazovaya promyshlennost no. 4, April 1970.

68. Bokserman, Yu. I., "Razvitie gazovoi promyshlennosti SSSR" (Development of the gas industry of the USSR), Neftyanik no. 2, February 1970: 9.

69. Energeticheskie resursy, op. cit.: 479.

70. Yufin, op. cit.: 21-22; Stroitelstvo truboprovodov no. 1, January 1972: 2.

71. Stroitelstvo truboprovodov no. 4, April 1970: 9.

72. Stroitelstvo truboprovodov no. 1, January 1970: 9; Sotsialisticheskaya Industriya, 28 March 1973: 1; Pravda, 15 November 1972: 3.

73. Stroitelstvo truboprovodov no. 4, April 1970: 10.

74. Stroitelstvo truboprovodov no. 1, January 1971: 2; Sotsialisticheskaya Industriya, 2 March 1973: 2.

75. Smirnov, K. K., "Masshtaby i zadachi truboprovodnykh rabot" (Scale and problems of pipeline operations), Stroitelstvo truboprovodov no. 4, April 1970: 8; Stroitelstvo truboprovodov no. 1, January 1971: 1; Stroitelstvo truboprovodov no. 12, December 1972: 8, 20.

76. Keller, A. A., Kornchenko, N. V., "Sotrudnichestvo stranchlenov SEV v oblasti neftyanoi i gazovoi promyshlennosti" (Cooperation of member countries of the CMEA in the oil and gas industries), Stroitelstvo truboprovodov no. 5, May 1970: 8; Grinyuk, V., "Magistral," Pravda, 6 March 1972: 3; Stroitelstvo truboprovodov no. 12, December 1972: 32; Izvestiya, 3 February 1973: 4.

77. Derezhev, S. P., "Mezhdunarodnye svyazy ministerstva gasovoi promyshlennosti" (International connections of the ministry

of the gas industry), Gazovaya promyshlennost no. 7, July 1970: 1; Pravda, 24 April 1971: 4; Time, 24 April 1972: 43; Ekonomicheskaya gazeta no. 17, April 1972: 20; Pravda, 27 March 1973: 4.

78. Sunday Times, 14 September 1969; UN Yearbook of International Trade Statistics 1968, New York 1970: 863: Sotsialisticheskaya industriya, 31 January 1971: 1; Ekonomicheskaya gazeta no. 21, May 1972: 21.

79. Madzhud, A. K., "Prirodnyi gaz Afganistana" (Afganistan natural gas), Gazovaya promyshlennost no. 7, July 1970: 5-6; Times, 24 November 1970: 7; Derezhev, S. P. "Mezhdunarodnye svyazy ministerstva gazovoi promyshlennosti," Gazovaya promyshlennost no. 7, July 1970: 2; Vneshnyaya Torgovlya SSSR za 1972 god, Moscow 1973: 100.

80. Ibid.: 2; Ekonomicheskaya gazeta no. 45, November 1970: 2.

81. Energeticheskie resursy, op. cit., 501-502.

82. Arbuzov, I. V., et al., "Sozdanie gazokhranilishch v vodonosnykh plastakh" (Creation of gas reservoirs in water-bearing layers), Stroitelstvo truboprovodov no. 2, February 1971: 29.

83. Ibid.: 26; Gazovaya promyshlennost no. 1, January 1971: 2; Pravda, 21 December 1971: 1.

84. Energeticheskie resursy, op. cit., 502-503.

85. Smirnov, V. A., "Razvitie edinoi sistemy gazosnabzheniya v SSSR" (Development of a single gas supply system in the USSR), Gazovaya promyshlennost no. 4, April 1970: 35.

86. "Pyatiletka stroitelei podvodnykh truboprovodov" (Five Year Plan of underwater pipeline builders), Stroitelstvo truboprovodov no. 1, January 1971.

87. Yufin, op. cit.: 53-54.

88. Aleksandrov, A. V., "Osnovnye tendentsii nauchnotekhnicheskogo progressa v gazovoi promyshlennosti," Gazovaya promyshlennost no. 4, April 1970: 28; Gudkov, S. F., "Nauchno-tekhnicheskii progress v gazovoi promyshlennosti" (Scientific and technological progress in the gas industry), Gazovaya promyshlennost no. 3, March 1972: 1-3.

Although natural gas is planned to become the main source of energy in the USSR by the end of the century, the primary fuel at present is petroleum. The share of oil in the fuel balance was 41 percent in 1970 and is expected to be over 44 percent by 1975.

The production and consumption of the world's oil is of vital importance in international relations, and it is frequently a decisive factor in determining the policy one nation adopts toward another. If domestic production of petroleum cannot keep pace with its rapidly rising requirements, the USSR will have to compete with the industrial nations of the West for oil from the developing countries. This could only lead to increased political tension in these areas, which are already dangerously prone to conflict.

HISTORICAL BACKGROUND

The early history of oil utilization in the area that is now the USSR is associated mainly with the Baku region of Azerbaidzhan. In the thirteenth century Marco Polo mentioned the export of naphtha on camels, and the origins of the Soviet petroleum industry were further described by subsequent travelers to the Caspian oil fields. In 1636, for example, the German Olearius saw the gathering and selling of petroleum, then used both as a medicine and as lamp fuel. He noticed thirty productive pits "within the range of one rifle shot."

Some recent data have been added to this chapter from the BBC monitoring service, USSR Weekly Economic Reports (Oil and Gas), 1972-73.

Russian interest increased in the 18th century, when expeditions from the St. Petersburg Imperial Academy of Science reported on the phénomenon of Baku petroleum. In 1813, after nine years of war, a treaty was signed with Persia that brought the khanates of Baku, Kuba, and Derbent under Russian sovereignty, thus extending Russia's control over most of the oil wealth of the Caucasus.

More systematic surveys were then initiated. Oil was being extracted by a very primitive method. Rags, ladles and buckets were dipped into the wells, which were generally less than 2 m. deep, though occasionally as deep as 20 m. In 1825 the Baku region had 125 wells, which produced 3,458 tons of oil. The average depth was 11.7 m. (maximum 29.5 m.) and the average diameter was 60 cm. Oil was extracted by hand-winch or by horse-powered pulleys.

It is claimed in Soviet sources that A. F. Semenov first suggested drilling for oil, using the same method that was then in use for fresh water extraction, and that the first well was actually drilled in 1848 at Bibieibat on the Apsheron peninsula. (Drilling did not begin in the Pennsylvania fields in the United States until 1859.) It was not until 1864, however, that manual drills began to be replaced by elementary steam-powered ram drills.

In 1870 at Mirzoev, prospectors drilling at a depth of 45 m. produced a gusher that marked the real beginning of Russian oil extraction on a significant scale.

From 1821 to 1871 some 361,200 tons of crude petroleum were extracted, of which 317,000 tons were from the Baku fields.

There was much useful research on the nature of petroleum and on the geological structures in which it was found; particularly well known are the series of publications by D.I. Mendeleev, which appeared after 1880.

Yearly production increased at a remarkable rate. From 1882 to 1916 over 210 million tons were produced. In 1901 alone output reached 11,500,000 tons (of which 10,980,000 tons was from Azerbaidzhan).

The explanation of this rapid growth is, first, that geological conditions were favorable, with large deposits near the surface, and second, that the route through the Black Sea to the consumers of industrial Western Europe compared well with the alternative of shipping supplies from the United States. Much foreign investment was attracted, mainly from Britain and the United States. In 1873 there were 12 oil firms in Baku; ten years later there were 79, and by 1903 there were 158. By the end of the boom, in 1913, the number of firms had reached 182. About 60 percent of the oil industry in Russia was then under the control of foreign investors.

New equipment was installed, and exploration and extraction techniques were improved. The average depth of oil wells in 1873 was under 50 meters; by 1900 it was about 300 meters.

Yet expansion was not constant; oil output dropped, and in 1905, the year of revolution, it was less than 8 million tons. By 1913, when 9,234,000 tons were produced, the level attained in 1901 had still not been surpassed.[1]

The development of the oil industry in the Caucasus was interrupted not only by strikes but also by the First World War and the subsequent revolution and civil war. Capitalist development in a typical 19th century manner had caused much labor unrest, which, added to political discontent, made Baku a revolutionary center. A strike on the oil fields in the summer of 1903 spread through the towns of the Caucasus and was only settled by the conclusion of a collective agreement between workers and employers. In 1917 the Bolshevik coup was given strong support by the workers in the oil industry. A Soviet regime was instigated in Baku in November 1917, and oil was nationalized in May 1918. During the years of Allied intervention in the Russian civil war, Britain showed an understandable reluctance to abandon the rich oil-bearing areas of the Caucasus; but the drop in production that occurred was caused mainly by the loss of markets and by fighting among the various factions and nationalities of the area. The occupation of Azerbaidzhan by the Red Army in April 1920 ensured that the future of the Caucasian oil industry would be in Soviet hands.[2]

Production continued to fall, however. In October 1920 some 282,000 tons of oil were extracted, but by April 1921 the monthly rate had dropped to 177,000 tons. Drilling dropped from a prewar average of about 10,000 m. per month to a mere 81 m. in November 1921. In Soviet histories of the oil industry the British have rather unfairly become scapegoats for this depressed period, although in fact they had left in the summer of 1919.

The new Soviet government did everything in its power to increase the output of oil, and in December 1921 about 246,000 tons were produced. Out of the total 1921 output for the USSR of 3,781,000 tons, Azerbaidzhan contributed about 2,500,000.

Petroleum was necessary not only for domestic consumption but also for exporting, in order to earn the hard foreign currency that was needed to buy vital new machinery abroad. Yearly output climbed slowly upwards, and by 1928 it reached 11,625,000 tons, somewhat above the prerevolutionary level of 1913. Azerbaidzhan and the smaller fields in the North Caucasus around Groznyi (which had been producing oil since 1821) could not ensure a sufficient rate of expansion, and prospecting was extended to other regions.[3]

The theoretical calculations about the geology of oil-bearing strata in the European part of the USSR that had been made by such specialists as I. M. Gubkin helped prospectors considerably in their search, and oil was found in 1929 near the river Chusovaya in the

Urals. The discovery of the Ishimbai field (Bashkir ASSR) in 1932
marked the beginning of a new era in the Soviet oil industry. The
Volga-Ural oil fields, which became known as the "Second Baku,"
were actually to overtake the original area in production.

Naturally, the Soviet government encouraged the expansion of
the oil industry as much as was possible with the available resources.
On 15 November 1930 the Central Committee of the Communist Party
decreed the further intensification of geological exploration and prospecting for oil, both in new regions and in the deeper horizons of
the old regions. The Central Committee directed that the search
for oil be increased to an extent that would ensure the discovery of
sufficient reserves to guarantee a yearly output of at least 15 million
tons by 1933. Special attention was to be paid to promising eastern
parts of the USSR, such as the Volga-Ural area, Emba area, and
Sakhalin, and other potentially rich regions, such as around Maikop,
Benoi, and Neftechala. These were to have first priority in trained
oilmen, materials, and equipment.

In the period from 1928 to 1932 output almost doubled, reaching
21,414,000 tons in 1932. The wells around Groznyi increased in
importance, producing over 8 million tons in 1931.

From 1932 to 1952 the oil industry stayed far below the coal
industry in its rate of development because there were not enough
known reserves to justify large-scale investment; in the USSR extraction of oil actually increased more slowly than in the world as a whole.

Even though the fuel industries were unable to fully satisfy
domestic requirements in the 1930s, when thousands of tractors were
being produced, creating a new demand for petrol, the USSR needed
foreign currency so urgently that oil was exported in large quantities.
As much as 15 percent of the total Western European market was at
one time being supplied by the USSR.

Resolutions at the 17th Congress of the Soviet Communist Party
in 1934 called for the creation of new industrial bases in the eastern
regions, and it was decided to construct several new petrochemical
plants (46 pipe stills and 93 petroleum cracking plants) and to lay
over 4,000 km. of pipeline for oil and oil products.[4]

As far back as 1863 the necessity for pipelines to transport
oil had been discussed by Mendeleev, but the first Russian pipeline
had not been built until 1878, when pipe of 76 mm. diameter was used
for a line of 8 km. near Baku. By 1883 there were 84 km., extending
to 280 km. by 1897. In 1906 the Transcaucasian Railway Company
built a kerosine pipeline from Baku over 883 km. to Batumi, with
pipe of 203 mm. diameter that was capable of transporting over one
million tons per annum. In 1911 an oil pipeline was laid 110 km.
from Shirvanskaya to Ekaterinodar (now Krasnodar) and a year later
from Shirvanskaya to Tuapse over 103 km. In 1914 Petrovsk was

connected to the Groznyi oil fields over a distance of 165 km., and
another pipeline was built to deliver oil from Dossor on the Emba
96 km. to the port of Bolshaya Rakusha, 16 km. of which were laid
on the bed of the Caspian Sea. This pipeline construction, which took
place before the Revolution, totaled 1,357 km.

During the first years after the revolution difficulty was experienced because of a shortage of imported pipe. In 1920 a proposed
pipeline from the Emba to Saratov, in which Lenin himself showed
keen interest, had to be abandoned for this reason.

After 1928 pipeline construction was continued, with pipes of
250 mm. diameter being laid 618 km. from Groznyi to a new refinery
at Tuapse and 823 km. from Baku to another new refinery at Batumi.
The working pressure of the Groznyi-Tuapse pipeline was 50 atmospheres. The Transcaucasian kerosene pipe was converted to oil.
In 1932 the first oil-product line in the USSR was completed over the
880 km. from Groznyi to Trudovaya of 300 mm. diameter pipe. In
1936, four years after oil was discovered at Ishimbai, in the new
Volga-Ural area, a 300 mm. diameter pipeline was laid to transport
oil over 168 km. to Ufa.[5]

At the 18th Congress of the Communist Party in March 1939
it was resolved that a new oil base should be created in the region
between the Urals and the Volga, a "Second Baku." This was not
only a sensible measure for the future economic growth of the USSR
but was also vital to the nation's security, in view of the imminent
possibility of invasion from Germany.

Petrochemical plants with a yearly capacity of 15 million tons
and cracking plants with 4.5 million tons were planned. The production of high-octane fuel and high-grade oils in particular was to
be increased beyond the Urals.

Plans, however, are not always fulfilled. Enthusiasm is no
substitute for specialized knowledge, and optimism can neither guarantee the discovery of oil in poorly explored areas nor immediately
overcome a shortage of metal for constructing cracking plants. For
instance, in January 1932 a target for 1937 of 80 to 90 million tons
of oil had been proposed, which had had to be modified in the actual
plan figures to 46.8 million tons. Actual production in 1937 was only
28,501,000 tons.

In 1939 the Volga-Ural area was developing much too slowly to
enable the oil industry to meet its planned targets. Drilling was
delayed by faulty equipment and poor working conditions. Although
work norms had been increased by over 25 percent in 1936, they could
not always be fulfilled because of the physical hardships being experienced by the oilmen.

In spite of the new wells in Turkmenia and the Volga-Ural area,
by 1940, when 31,121,000 tons were extracted in the USSR, some

87 percent was still from the Caucasus; Azerbaidzhan alone contributed 75 percent of the total. The Volga-Urals oil fields produced under 6 percent and Central Asia and Kazakhstan under 5 percent.[6]

The war forced the government to further intensify efforts to expand the extraction and processing of oil in the eastern regions, which by the end of 1942 were contributing 18.3 percent of the total production. Because of the fear of enemy occupation, output at the Groznyi fields was curtailed and much of the equipment dispatched to the east. In 1943 some 4 million tons less oil was produced than in the previous year, and in Baku output had to be limited because fighting on the Volga and in the northern Caucasus prevented the delivery of oil to the Center. In 1941 over 23 million tons were extracted in Azerbaidzhan; by 1945 this had dropped to 11 million tons.

The attempt to expand the oil industry in the Volga-Ural area, Kazakhstan, and Uzbekistan was reasonably successful. Despite the drop in the extraction of crude petroleum in 1943, the production of petrol rose to 110 percent, of diesel oil to 270 percent, and of motor oil to 170 percent of the 1942 figures.[7]

It has been calculated that from July 1941 to July 1945 over 91 million tons of oil were extracted in the USSR, a yearly average of 22.8 million tons. Compared with this, the shipment of 2,670,000 tons of petroleum products by the Western allies to the Soviet Union during the war does not at first seem significant. Yet because this was composed largely of high-octane petrol for Soviet fighter and bomber aircraft, and since substantial quantities of advanced oil refinery equipment were also supplied, the importance of Anglo-American aid should not be overlooked.[8]

Although after the war the oil industry continued to expand in the Caucasus at Baku, Groznyi, and Krasnodar and in the Ukraine, the contribution of the eastern regions grew steadily in proportion. The discovery of such important fields as Tuimazinskoe, Serafimovskoe, Aleksandrovskoe, Bavlinskoe, and Zolnenskoe helped to raise the share of the eastern regions in the total Soviet output from 12 percent in 1940 to 44 percent in 1950.

Until 1950, when 37,878,000 tons were extracted, the rate of growth remained comparatively low. The fifth Five Year Plan, however, called for an 85 percent increase in oil output in the period from 1951 to 1955. The amount of exploratory drilling was expanded, and improved technology helped to boost production.

By 1955 output had climbed rapidly to 70,793,000 tons, but the Soviet fuel balance remained much as before: coal provided 64.8 percent, while oil provided 21.1 percent and gas only 2.4 percent. In the United States at that time the more economical fuels, oil and gas, formed 64.4 percent of the fuel balance.

After 1956 the importance of developing the oil and gas industries was fully recognized, and their share in the nation's capital investment was greatly improved. Production of oil has since risen very rapidly indeed, with the yearly increase alone averaging about 20 million tons. (See Table 3.1 and Figure 3.1.) By 1970 oil and gas made up 60 percent of the Soviet fuel balance, of which oil was 41 percent, having surpassed coal as the primary source of power in 1968.[9]

The geographical distribution of production had also changed greatly over this period. By 1965 the Volga-Ural fields had become the main producing area, supplying about 72 percent, while the share of the Caucasus had dropped to 18 percent. In the late 1960s production in the new oil fields of Mangyshlak in western Kazakhstan and in Tyumen in West Siberia had increased greatly.[10]

Pipeline construction since the war has kept pace with the growth in oil output. During the war fuel was supplied to the besieged city of Leningrad by a temporary petrol pipeline laid over the bed of Lake Ladoga. Pipe of 100 mm. diameter was used, and the pipeline, which was built in only fifty days, stretched over 31 km., about 21 km. of which was under water. When the battle of Stalingrad closed the Volga to oil barges a pipeline was constructed from Astrakhan to Saratov, a distance of 655 km., in just eight months. In 1944 a pipeline from Okha in Sakhalin to Komsomolsk-na-Amure was completed, stretching over 650 km. through the ravines and marshes of the Far East.

By 1955 there were 11,000 km. of major oil pipeline in the USSR. In 1957 construction began on the Trans-Siberian pipeline, which was to extend over 3,682 km. from Tuimazy through Omsk to Irkutsk. Large 720 mm. diameter pipe was used; the first stretch to Omsk went into operation in 1959, and the remaining section to Irkutsk was completed by 1964. By 1972 the oil pipeline network stretched over 40,000 km.[11]

Pipeline construction in the 1960s and 1970s, such as the Druzhba (Friendship) system to the Comecon countries and the Siberian network, is dealt with in greater detail below, in the section on the transporting of oil.

Recent developments in the exploration and extraction of oil in the Soviet Union are also examined later in this chapter, and the utilization of oil in industry is discussed in Chapter 8.

PRESENT AND FUTURE DEVELOPMENT

Exploration and Location of Reserves

In the nineteenth century the search for oil was based on the uncertain evidence of surface seepage, release of gases, and signs of bitumen-impregnated sand or limestone.

TABLE 3.1

Production of Oil in USSR,* 1901-2000
(in millions of metric tons)

Year	Output	Year	Output
1901*	11.5	1962	186.2
1907	8.7	1963	206.1
1913	9.6	1964	223.6
1913	10.3	1965	242.9
1917	8.8	1966	265.1
1921	3.8	1967	288.1
1928	11.6	1968	309.2
1932	21.4	1969	328.3
1937	28.5	1970	352.7
1940	31.1	1971	372.0
1945	19.4	1972	394.0
1946	21.7	1973	429.0
1950	37.9	1974	461.0
1955	70.8	1975	496.0
1960	147.9	1980	700.0
1961	166.1	2000	1000.0

*1901-1913, Russian Empire; 1913-1972, present borders of the USSR; 1973-2000, planned.

Sources: Narodnoe khozyaistvo SSSR v 1969 godu, Moscow 1970; Pravda, 21 December 1971: 3; Ekonomicheskaya gazeta no. 5, January 1972: 4; Pravda, 30 January 1973: 1; Pravda, 26 January 1974: 2.

Virgin areas are now submitted to careful aerial reconnaissance, when photographs are taken to reveal potential oil-trap structure. Favorable areas are then prospected by geological methods. Gravimetric, electrical, magnetic, and seismic methods have been used for several years in the USSR, and recently surveys based on sonic, radioactive, and electromagnetic testing have given successful results. The potential productivity of an oil-bearing area is then estimated by analyzing rock specimens from core drilling before a promising site for an oil well is selected. As with gas reserves, oil deposits are classified from predicted to proved, passing through the categories from D_2 to A according to the intensity of exploration.

FIGURE 3.1

Expansion of Oil Production in USSR, 1950-1973

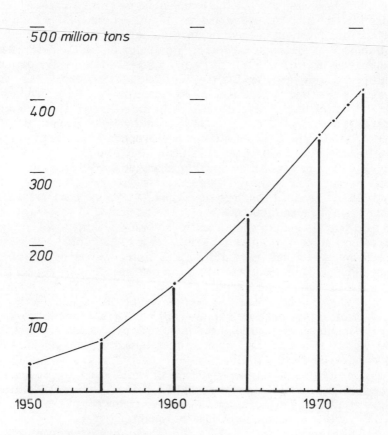

Source: Table 3.1.

Drilling is the most expensive operation in the process of seeking or extracting oil, requiring over half the capital investment in the oil industry.

In the earliest period percussion or cable-tool methods were used, but ramming was gradually replaced by rotary drilling, which more than doubled the speed and the depths attainable. In 1925 a new method was invented in Baku by an engineer, M. A. Kapelyushnikov, who thought of placing a turbine in a cylinder below the drill pipe to convert the hydraulic power of the mud stream into rotary power. It was modified by other specialists until the new turbodrill was capable of speeds over ten times those of rotary drills.

Used for only 2 percent of Soviet oil drilling in 1940, the turbodrill accounted for as much as 87 percent by 1960. It is still the main method in the USSR, though its use had dropped to 80 percent in 1965 and 70 percent in 1970, since it is unsuitable at depths of over 3,000 m. A method has been developed of changing the bit in about 70 minutes, without raising the drill pipes to the surface.

An electrodrill has been developed in recent years for drilling to depths of around 5,000 m., but by the beginning of the seventies several complicated technological problems had still to be overcome, including that of poor-quality Soviet steel.

The most widely used drilling rig is the Uralmash 3 D-50, which is designed to drill to depths of up to 5,000 m. An experimental rig, the BU-315, has reached depths of 7,000 m.

Diamond-faced bits have become widely used; in 1964 they were tried in 22 oil regions, with an average saving of about 45 rubles per carat consumed, compared with the cost of using ordinary bits for drilling hard rock at great depths, even though a diamond-headed bit costs between 2,000 and 4,000 rubles.

The drilling of narrower diameter wells has also proved economical by saving metal and cement, and by 1970 half the wells drilled (37 percent of the exploratory wells and 61 percent of the development wells) were less than nine inches in diameter.

The transport and erection of derricks has been accelerated, using powerful caterpillar tractors, either without dismantling the rigs for short distances or by moving them in about twelve sections for longer stretches. Interesting experiments have been carried out using hovercraft to transport derricks in the marshes and snow of the Far North.[12]

Improvements in technology have greatly increased the amount, depth, and speed of drilling that is being achieved in the USSR, as may be seen in Table 3.2.

In the fifty years from 1871 to 1920 less than 4 million m. of exploratory and development wells were drilled, while in the five years from 1961 to 1965 over 47 million m. were drilled. In the same

TABLE 3.2

Improvements in Drilling, 1930-69

Indices	1930	1940	1950	1960	1969
Amount of drilling (in thousands of meters)	665	1,830	4,283	7,715	10,967
Average depth of wells (in meters)					
Development	664	1,002	1,148	1,586	1,683
Exploration			1,362	1,928	2,446
Monthly drilling average per rig (in meters)					
Development	104	412	629	993	1,121
Exploratory	65	233	209	401	339

Source: Melnikov, N. V., Mineralnoe Toplivo, Moscow 1971: 47.

five-year period in the United States, by way of comparison, over 284 million m. were drilled.

Academician Trofimuk, however, has complained of poor productivity. Over half the wells drilled prove to be dry. Some 90 percent of the prospecting meterage drilled gives poor results, possibly because prospecting organizations are paid for the number of meters drilled, rather than for the increases in known reserves that result.

The geographical distribution of exploratory drilling has changed greatly in the Soviet period, as may be seen in Table 3.3. From 1920 to 1940 most prospecting, 71 percent, was carried out in Azerbaidzhan and in the North Caucasus, with only 8 percent in the Volga-Urals. In the following decade test drilling was distributed more evenly between Azerbaidzhan, the North Caucasus, and the Volga-Urals. Since 1951 exploration has been concentrated in the Volga-Urals, with notable increases in the Ukraine, Siberia, and the Komi ASSR. Exploration has dropped sharply in Azerbaidzhan and in the North Caucasus.[13]

Almost 11.2 million sq.km. of Soviet territory may contain deposits of oil. A further 4 million sq.km. of potential oil-bearing area is situated on the continental shelf. Of this potential area, only Azerbaidzhan and the Checheno-Ingush ASSR have been subjected to very intensive exploration (over 250 m. drilled for every square kilometer of potential oil-bearing area). The largest potential areas are in Siberia, the Komi ASSR, Kazakhstan, and Turkmenia, yet in these

TABLE 3.3

Distribution of Test Drilling, 1920-70
(major areas as percentage of USSR total)

Area	1920-40	1941-50	1951-60	1961-70
Volga-Urals	8	26	40	34
North Caucasus	29	23	20	15
Komi ASSR	2	3	2	5
Siberia	—	—	2	10
Azerbaidzhan	42	25	11	5
Ukraine	1	5	7	11
Kazakhstan	7	6	4	4
Central Asia	4	8	9	10
Others	7	4	5	6

Sources: Energeticheskie resursy SSSR, Toplivno-energetiches- kie resursy, Moscow 1968: 277-80; Neftyanoe khozyaistvo no. 3: 1971.

areas the density of exploratory drilling is under 10 m. per square kilometer of potential oil-bearing area. In 1970 the average for the whole USSR was also below 10 m. per square kilometer, and it would therefore be unwise to attempt to give definitive figures for Soviet oil reserves, even if the government were willing to allow more recent data on this subject to be published.

The first calculation of oil reserves in the USSR was made by I. M. Gubkin in 1924. He estimated that the total reserves (A + B + C_1) were 2,882 million tons, of which 50 percent were in Azerbaidzhan, 31 percent in the Groznyi area of the North Caucasus, and only 9 percent in the Ural-Emba region. These calculations gave the USSR greater reserves than any other nation, 37 percent of the world total.

By 1937 increased prospecting had allowed Gubkin to raise his estimate to a total of 6,376 million tons (A + B + C_1), which were distributed mainly in Azerbaidzhan (49 percent), the Emba area (17 percent), and the Volga-Ural area (12 percent).

As for present reserves, the United Nations Yearbook gives figures for 1969 of 73,062 million tons for the world total, 7,987 million for the USSR, and 4,004 million for the United States. This would suggest that the USSR has about a ninth part of the world's known reserves. However, this is unlikely to be an accurate indication of the present position, since Soviet oil reserves have been a state secret since 1947. In fact, the USSR claims 37 percent of the earth's

potential oil-bearing areas. Working from estimates that the average
density for the potential areas of the world is 16,400 tons of oil per
square kilometer would give the USSR, with 11,900,000 sq.km. of oil-
bearing territory, about 195,160 million tons of ultimate oil reserves.
This compares well with figures for the ultimate reserves of the
United States of around 100,000 million tons, although these are more
densely contained, within an area of less than 5 million sq.km.[14]

The proved oil reserves of the Soviet Union (categories A + B +
C_1) are situated at the following depths: 15 percent at less than 1,200
m.; 66.4 percent at 1,201 to 1,800 m.; 11.2 percent at 1,801 to 2,400
m.; 5.2 percent at 2,401 to 3,000 m.; 4.3 percent at more than 3,000
m. Since over 80 percent of proved oil reserves lie at depths of less
than 2,000 m., the possibility of extraction is very favorable. In 1961
over 60 percent of the oil reserves was concentrated in just 21 large
oil fields, which has greatly simplified the process of extraction.

Potential crude petroleum reserves are divided into those with
low sulphur content (under .5 percent), medium sulphur content (.5 to
2 percent), and high sulphur content (up to 5.5 percent); Soviet reserves
are 24 percent low sulphur, 52 percent medium sulphur, and 24 percent
high sulphur crudes. The low sulphur crudes are mainly in Azerbaidzhan, Central Asia, and the Ukraine.[15]

During the eighth Five Year Plan (1966-70) proved reserves were
expanded in West Siberia, the North Caucasus, Komi ASSR, Udmurt
ASSR, Perm oblast, Orenburg oblast, the Ukraine, and Turkmenia.

A representative of the State Planning Commission (GOSPLAN
SSSR) nonetheless had to criticize prospecting organizations for their
inefficient geological exploration over this period. The planned expansion of reserves in the three categories A + B + C_1 was under-
fulfilled by 10.9 percent, and the two top categories alone (A + B)
were 16.2 percent below target.[16]

The exploration of the world's ocean beds has been increasing
in recent years, and it is possible that by 1980 some 35 percent of
total oil extraction may be from offshore wells. A Soviet estimate in
1971 of ultimate oil reserves at accessible depths under the sea was
as high as 135,000 million tons.[17]

Oil has been extracted from the bed of the Caspian Sea for many
years, but this has been almost completely from the western shores
off Azerbaidzhan. The Turkmenian shores are also rich in oil, and
prospecting is being intensified in eastern parts of the Caspian. A
new oil port is being built on the Cheleken peninsula.

There is geological evidence that the Arctic Ocean may prove
to be as rich as the Gulf of Mexico, and prospecting has already begun.
The first exploratory well was sunk in 1972 in the area of Kolguev
Island, although long delays were caused by storms. In December
1971 a well 4,200 m. deep was drilled by an oil survey expedition in
the Pechora estuary north of Arkhangelsk.

Oil has been discovered in the Okhotsk Sea off the coast of Sakhalin. In December 1971 one well was producing 100 tons daily. The further exploration of the northeast coast is planned.[18]

The known oil reserves $(A + B + C_1)$ are situated mainly in the Volga-Ural area; there are also substantial reserves in the North Caucasus and Siberia. The ultimate reserves, however, lie mainly in Siberia. It is therefore expected that the intensive exploration for oil that has begun in Siberia will continue through the 1970s and that the main increase in production will be from Siberian fields, especially those in Tyumen oblast and the Yakut ASSR. In 1972 several geological expeditions were prospecting for oil in the area between the Enisei and the Lena, and in the depression between the Kolyma and the Indigirka rivers. There are 3.2 million sq.km. of potential oil-bearing areas in Eastern Siberia, which is 50 percent more than in Western Siberia, but by 1972 only 400 wells had been drilled, forming an average of .3 m. per square kilometer, compared with 3.5 m. in Western Siberia, 57 m. in Tataria, and 82 m. in Kuibyshev oblast. V. D. Shashin, the Minister of the Oil Industry, insists that the 470,000 m. being drilled in the years 1972-75 is not sufficient and that a yearly rate of over one million m. must soon be attained to allow production to meet requirements in 1990.[19]

Potentially rich oil-bearing areas are shown in Map 3.1. They correspond very closely to the regions already described as having large reserves of natural gas. As with gas, known reserves of petroleum in the USSR are being steadily expanded by constant exploration. While the abundant data that are available for natural gas are not available for oil, two points in particular arise repeatedly in Soviet sources, first, that the ratio of consumption to reserves in the USSR is very favorable, especially compared with the situation in the United States, and second, that Soviet oil experts are sufficiently optimistic about steadily expanding reserves to predict that the present rapid acceleration in output will continue over several decades to come.

Production Statistics

In 1971 the total world production of crude oil was 2,465 million metric tons. The major oil-producing nation was the United States, with 474 million tons, and the USSR came second with 372 million tons. The other largest producers of oil are shown in Table 3.4

As mentioned above, the Soviet Union is in a stronger position than the United States as regards domestic oil reserves. The rate of extraction in the USSR has been rising steadily over recent years at about 7 percent, compared with the world average increase of 5.5 percent; there was actually a slight drop in production in the United

TABLE 3.4

Output of Major Oil-Producing Nations, 1969-71
(in millions of metric tons)

Country	1969	1970	1971
World	2076	2272	2399
USA	456	475	467
USSR	328	353	372
Iran	168	192	224
Saudi Arabia	149	177	223
Venezuela	188	194	186
Kuwait	130	137	147
Libya	150	162	132
Iraq	75	77	84
Canada	54	61	64
Nigeria	27	54	76
United Arab Emirates	29	38	52
Algeria	45	48	38

Source: United Nations Statistical Yearbook 1972, New York 1973: 180-181.

States in 1971. Even allowing for factors such as conservation during the world glut of oil, output of oil in the United States is unlikely to expand at the same speed as in the USSR, which will almost certainly be the major producer by the 1980s. The comparative production of oil in the USSR and the United States since 1960 is shown in Table 3.5 and Figure 3.2.

Production per capita will remain higher in the United States far longer, however, in view of its smaller population, and it is also worth noting that while the USSR exports large quantities of its crude oil, the United States refines and consumes far more than it produces. In 1970, for example, indigenous output in the United States was responsible for only 73.6 percent of its total oil consumption of 645 million tons; the remainder was imported mainly from South America (16.4 percent) and Canada (5.7 percent). A mere 2.4 percent came from Africa and the Middle East, but in view of the threatening energy gap, the United States will have to increase imports from this area considerably.[20] Soviet trade in oil is discussed in Chapter 8.

Legend

A   Lithuania and Kaliningrad Oblast

B   Belorussia

C   Ukraine

D   North Caucasus

E   Azerbaidzhan

F   Volga-Ural

G   Komi ASSR

H   Western Siberia

I   Eastern Siberia

J   Sakhalin

K   Kazakhstan

L   Turkmenia

M   Uzbekistan

MAP 3.1

Oil-Bearing Areas

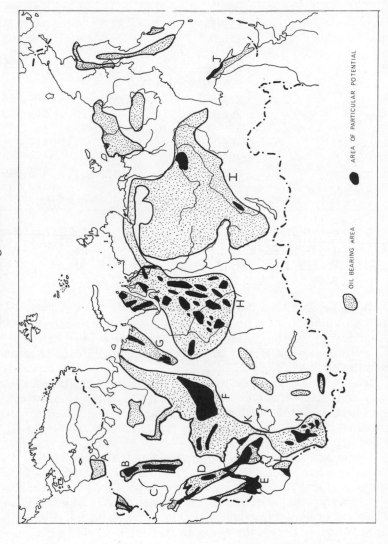

Sources: Energeticheskie resursy SSSR—Toplivo-energeticheskie resursy, Moscow 1968; Atlas razvitiya khozyaistva i kultury SSSR, Moscow 1967, 10–11; Sovetskii Soyuz—Obshchii obzor, Moscow 1972, 486–87.

○ OIL BEARING AREA

● AREA OF PARTICULAR POTENTIAL

TABLE 3.5

Comparative Oil Production, USSR and
United States, 1960-71
(in thousands of tons)

Year	USSR	United States
1960	147,859	347,975
1961	166,068	354,303
1962	186,244	361,658
1963	206,069	372,001
1964	223,603	376,609
1965	242,888	384,946
1966	265,125	409,170
1967	288,068	434,705
1968	309,150	449,885
1969	328,299	455,656
1970	353,039	475,289
1971	372,000	466,704

Sources: United Nations Statistical Yearbook 1970, New York 1971: 213; SSSR v tsifrakh v 1971 godu, Moscow 1972: 60-61; United Nations Statistical Yearbook 1972, New York 1973: 181.

By the end of 1965 the amount of crude oil that had been extracted in the USSR since the oil fields were first exploited reached 2,771 million tons; in the five following years alone well over half this total, 1,543 million tons, was produced. This accelerated rate of output has continued in the early 1970s.

It is certainly possible to show how original plan targets were not attained; in 1970, for example, when it was hoped to produce 390 million tons, only 353 million tons were actually extracted.[21] Yet an impressive growth in the oil industry has, in fact, been achieved, as may be seen in Table 3.1 and Figure 3.1.

The geographical distribution of production since 1940 is shown in Tables 3.6 and 3.7 and in Figure 3.3. The most important change in this period was the emergence of the Volga Urals over Azerbaidzhan as the major producing area. In 1940 Azerbaidzhan was contributing 71.5 percent of the total output, dropping steadily to 8.8 percent in 1965; in the same period the share of the Volga-Urals fields rose from 5.8 percent to 71.5 percent; this was almost a complete reversal of proportions. Trends in the 1970s are for both areas to drop in

FIGURE 3.2

Comparative Oil Production, USSR and United States

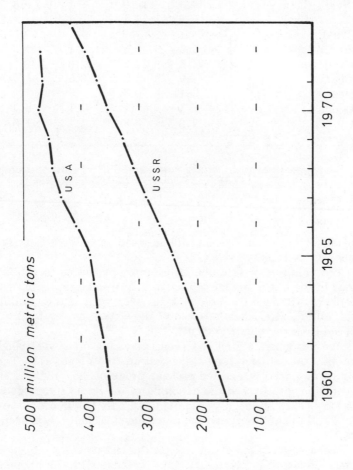

Source: Table 3.5.

TABLE 3.6

Production of Oil, 1940-70
(in percentage of USSR total)

Area	1940	1950	1960	1965	1970
RSFSR	22.4	48.2	80.4	82.3	80.6
Komi ASSR	0.2	1.3	0.5	0.9	1.5
North Caucasus	14.8	16.0	8.1	8.6	9.6
Volga-Ural	5.8	29.3	70.7	71.5	58.7
West Siberia	—	—	—	0.3	8.9
Far East	1.6	1.6	1.1	1.0	0.7
Azerbaidzhan	71.5	39.2	12.0	8.8	5.7
Turkmenia	1.9	5.3	3.6	3.9	4.1
Ukraine	1.1	0.8	1.5	3.1	3.9
Kazakhstan	2.3	2.7	1.1	0.9	3.7
Others	1.8	3.8	1.4	1.0	2.0

Sources: Energeticheskie resursy SSSR, Toplivno-energetiches-kie resursy, Moscow 1968: 322; SSSR i soyuznye respubliki v 1970 godu, Moscow 1971; Table 3.11; Table 3.12.

importance, while the share of the new fields in West Siberia and Kazakhstan is rapidly rising.

The relative importance of the main producing republics and regions in the development of the Soviet oil industry is indicated in Table 3.8. Although the Volga-Urals area went into exploitation more than a century after the oil fields of the Caucasus, by 1965 it had produced a larger quantity of oil.

Azerbaidzhan, which has been producing oil since 1808, reached a total of 1,000 million tons extracted on 28 March 1971. The oilmen of the Tatar ASSR, where the richest fields of the Volga-Ural area are concentrated, celebrated the attaining of the same figure soon after on 14 May 1971, although production had only begun after the war; indeed, half of this total had been produced in just the preceding five years.[22]

In the description of the oil-producing areas of the USSR that follows, emphasis is placed not only on those with the highest present output but also on those expected to contribute most to the future growth of the Soviet oil industry.

In Soviet sources distribution is normally presented by republic, but in the case of the RSFSR, which accounts for over 80 percent

FIGURE 3.3

Geographical Distribution of Oil Production, 1940-1975,
Percentage of Total Output

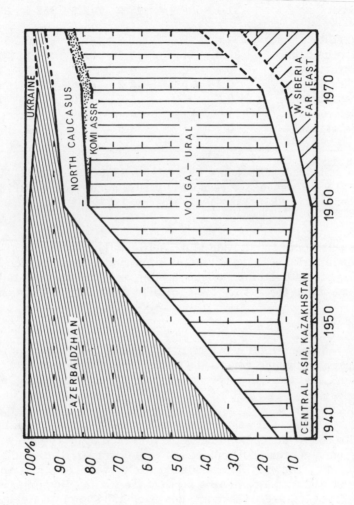

Sources: Tables 3.6 and 3.12.

TABLE 3.7

Production of Oil, by Republic, 1940-71
(in millions of tons)

Republic	1940	1950	1960	1965	1970	1971	1972
USSR	31.1	37.9	147.9	242.9	352.6	372.0	394.0
RSFSR	7.0	18.2	118.9	199.9	284.3	304.0	321.0
Azerbaidzhan	22.2	14.8	17.8	21.5	20.2	19.2	18.4
Turkmenia	0.6	2.0	5.3	9.6	14.5	15.5	15.9
Ukraine	0.4	0.3	2.2	7.6	13.9	14.3	14.5
Kazakhstan	0.7	1.1	1.6	2.0	13.2	16.0	18.0
Belorussia	—	—	—	—	4.2	5.3	5.8
Uzbekistan	0.1	1.3	1.6	1.8	1.8	1.8	1.4

Sources: Narodnoe khozyaistvo SSSR v 1969 godu, Moscow 1970: 197; SSSR i soyuznye respubliki v 1972 godu, Moscow 1973; Narodnoe khozyaistvo SSSR 1922-1972, Moscow 1972.

of petroleum output, this is not sufficiently informative. The central administration of the Soviet oil industry delegates responsibility to large territorial unions (obedinenie), and it is possible to break down total output figures for the USSR into those for the main unions.

RSFSR

The main oil province in the Russian Federation is the Volga-Ural area, which covers the Tatar ASSR, Bashkir ASSR, and Udmurt ASSR, and the Kuibyshev, Volgograd, Saratov, Orenburg, and Perm oblasts.

Tatar ASSR: Oil was not exploited in the Tatar ASSR until 1943, although evidence pointing to the presence of petroleum was reported as early as the 18th century and exploratory drilling had been started in 1939. The vast Romashkinskoe field, which occupies an oil trap some 4,000 sq.km. in area north of the town of Bugulma, was discovered in 1948. There are now over 120 known oil fields in the Tatar ASSR. Among the other most important fields are the Bablinskoe, 40 km. east of Bugulma; Bondyuzhskoe, in the basin of the river Kama; Novoelkhovskoe, east of the Romashkinskoe; and the Pervomaiskoe, east of the town of Menzelinsk.[23]

TABLE 3.8

Total Cumulative Output of Oil, by Region
(in millions of tons)

Region	Year Exploitation Began	Total Output by 1 January 1966	Output 1966-70	Total Output by 1 January 1971
USSR		2,771	1,543	4,314
RSFSR		1,650	1,254	2,904
Volga-Ural	1928	1,290	950	2,240
North Caucasus	1821	3,000	140	240
Azerbaidzhan	1808	888	105	993
Turkmenia	1873	86	64	150
Ukraine	1940	60	60	120
Kazakhstan	1911	40	39	79
Others		47	21	68

Sources: Energeticheskie resursy SSSR: Toplivno energeticheskie resursy, Moscow 1968: 323; 1966-70: Calculated from data in
text and from SSSR i soyuznye respubliki v 1970 godu, Moscow 1971,
and other statistical yearbooks.

From 1943 to 1965 over 340 million tons of oil were extracted in
the Tatar ASSR: the rate of growth was so rapid that by 14 May 1971
oilmen were able to erect a monument to celebrate the production of
the 1,000-millionth ton.* In 1970 output was 100 million tons, over
a quarter of the total for the USSR, and it is claimed that reserves
are sufficient to ensure that this level of extraction will be maintained
for decades to come. The same rate has certainly continued in 1971

---

*Much publicity is given in the USSR to the attaining of such
rounded numbers. Workers may be inspired to greater exertions to
reach these targets, and it is perhaps as good a reason for a cele-
bration (albeit on a smaller scale) as the centenary of Lenin's birth
or the fiftieth anniversary of the founding of the Soviet Union. At any
rate, such occasions usually provide the researcher with some interesting statistics!

and 1972, although its proportion to total USSR production is of course gradually dropping.

Efficiency in the "Tatneft" union has been high, and many technological innovations have been introduced. A new drilling bit was developed that bored an average length of 33.5 m. at a speed of 18.6 m. per hour. The first rigs to drill at a speed of 5,000 m. a month were operating in the Tatar ASSR. By 1970 the average time taken for derrick construction had dropped to 2.6 to 2.7 days, and the average period between repairs had lengthened to 227 days. Some 70 percent of the wells have been mechanized, and about half of them now work on a system of hermetically sealed extraction and transportation of oil and gas. This has helped to raise gas utilization to almost 90 percent.[24]

With the development of the Tatar oil industry in uninhabited areas has come the growth of several new towns, such as Almetevsk on the Romashkinskoe field, which is known as the "capital of the Tatar oilmen."[25]

Bashkir ASSR. Oil has been produced in Bashkiria since 1932, when the first gusher was discovered at Ishimbaevo. There are now over a hundred oil fields, of which the largest are the Tuimazinskoe, 180 km. west of Ufa; Shkapovskoe, 35 km. south of Belebei; and the Arlanskoe, 160 km. northwest of Ufa.[26]

By the fortieth anniversary of the Bashkir oil industry on 27 June 1972, over 670 million tons of oil and 35,000 million cu.m. of gas had been produced. Yet expectations have not been fulfilled. Instead of 60 to 70 million tons, in 1970 only 40.7 million tons of oil were extracted, actually less than in previous years (in 1968, for example, output was 44.4 million tons). The older fields, such as Tuimazinskoe and Shkapovskoe, are almost exhausted, and the "Bashneft" union has been criticized for the slow rate of mechanization in some of the newer wells. According to criticisms published in Soviet oil journals in 1972, labor discipline in Bashkiria has been weak, and there have been considerable losses due to accidents and breakdowns. Trade union committees were attacked for not improving working conditions and social amenities sufficiently. It is hoped, however, that by intensifying exploration Bashkiria can maintain its output in the 1970s at about 40 million tons a year. Great efforts are being made to raise the proportion of oil recovered from the layer from 40 percent to at least 55 percent; this has already been attained at the Tuimazinskoe field.[27]

Kuibyshev Oblast. Since the first oil was extracted in 1936 over a hundred fields have been discovered in the oblast. The largest are the Mikhailovsko-Kokhanovskoe, 22 km. southeast of the small town

of Tolkai; Kuleshovskoe, 140 km. southeast of Kuibyshev; and Dmit-
rievskoe, 30 km. south of Buguruslan.

From .2 million tons in 1940, output has risen to 35.4 million
in 1971, and it is hoped that this level will be maintained for several
years. On 28 June 1972 the extraction of their 500-millionth ton
earned Kuibyshev oil workers a front-page photograph in Pravda.

Production from flowing wells has dropped from 71 percent in
1965 to 55 percent in 1970 as the older fields approach exhaustion.
Exploration has been intensified, however, using more advanced
technology, and drilling speeds of 2,200 m. per rig per month have
been attained, allowing wells of 3,000 m. to be drilled in less than 45
days. In the period from 1971 to 1975 some 44 new fields are to be
brought into production, and the "Kuibyshevneft" union is optimistic
about maintaining output at the 1970 level.

Here too the oil industry has been responsible for the appear-
ance of new towns, such as Pokhvistnevo, Zhigulevsk, and Otradnyi.[28]

Volgograd Oblast. Prospecting began in 1940, with the first intensive
exploratory drilling in 1945. In 1951 oil was produced at the Korobkovskoe field northwest of Kamyshin. Most of the fields in Volgograd
oblast contain both oil and gas or are pure gas fields, and production
of oil is dropping from the 6 million tons extracted in 1965.[29]

Saratov Oblast. Exploration began in 1935, and about forty fields have
been brought into exploitation, although several produce more gas than
oil. The Sokolovogorskoe, near Saratov, is the principal oil field. A
total of about 30 million tons had been extracted in Saratov oblast by
1970.[30]

Orenburg Oblast. This area has greater potential than both Volgograd
and Saratov oblasts. Oil was first discovered at the Buguruslanskoe
field in 1937, and about thirty oil fields are now known. The Pono-
marevskoe, on the Dema river, was the most important of the earlier
discoveries. Exploration is being intensified in the eastern part of
the oblast, and from 1966 to 1970 reserves were significantly increased.
The Pokrovskoe, Bobrovskoe, and Sorochinsko-Nikolskoe fields, which
have considerable reserves, are now in production and supply much
of the yearly increase.

In 1965 output was 2.6 million tons, and by 1970 it had risen to
7.4 million; total output for 1966-70 was 26.9 million tons. By 1975
drilling is to be double the 1970 level, and there are now three drilling
enterprises in Orenburg oblast, where earlier there was only one.
Production is expected to continue to expand in this area at the rate
of about one million tons a year. Output reached 8.4 million tons in
1971, was over 9.4 million in 1972, and was expected to be 10.7 million
in 1973.[31]

Perm Oblast. The first oil in the Volga-Ural area was discovered in this oblast in 1929 at the Verkhne-Chusovskoe field. There are now some eighty oil and gas fields, among the biggest of which are the Krasnokamskoe, on the Kama river, 30 km. west of Perm; Polaz-ninskoe, on the Kama river, 40 km. north of Perm; and Yarinskoe-Kamennolozhskoe, between the Kama and Chusovaya rivers, northeast of Perm.

Drilling density varies from about 80 m. per square kilometer in the producing areas to less than 5 m. in some favorable parts of the Ural foothills; it is planned to drill 1.3 million m. of exploratory wells in the 1971-75 period, concentrating on the regions with the highest potential.

In 1970 about 16.1 million tons were extracted, bringing the total for the eighth Five Year Plan to 70.5 million tons, compared with 30.4 million tons in the 1961-65 period and the total for 1929-70 of 112 million tons. Although production from established fields is likely to drop below 7 million tons by 1975, it is still hoped that about 27 million tons will be extracted in 1975 because several new fields will be brought into exploitation.

Exploration is made more difficult by the huge Kama reservoir and by large areas of marsh and trackless taiga, but the "Permneft" union expects to increase oil output considerably in the 1970s.[32]

Udmurt ASSR. Although prospecting began in Udmurtia in 1945, it concentrated on the less favorable southern territory, and by 1968 only ten oil fields had been discovered. The Arkhangelskoe, Kien-gopskoe, and Gremikhinskoe were the first fields to be prepared for exploitation. Further exploration in the northern territory had raised the number of known fields to twenty by 1972, when over a million tons was extracted, and Udmurtia is expected to be producing 6 million tons per annum by 1975.[33]

Most of the oil extraction in the above regions of the Volga-Ural oil province is from Devonian and Carboniferous formations at depths between 1,500 and 2,500 m. The oils vary greatly in quality, but tend to be high-sulphur crudes with a low octane number, making refining expensive.

The period of rapid growth in the Volga-Ural area is probably at an end. The main producing areas, Tatar ASSR, Bashkir ASSR, and Kuibyshev oblast, are making intensive efforts simply to maintain their present high output. While increased output may be expected from the Orenburg and Perm oblasts and from the Udmurt ASSR, this will not be sufficient to enable the Volga-Ural area as a whole to keep pace with the average rate of development for the USSR. Its share in total production is therefore dropping steadily.

Northern Caucasus

The Northern Caucasian oil province is exploited by the oil unions
of the Chechen-Ingush ASSR, Dagestan ASSR, Krasnodar krai, and
Stavropol krai and is one of the longest-established oil-producing
areas.

Chechen-Ingush ASSR. Oil from pits was utilized in the Chechen-
Ingush ASSR even before the eighteenth century, but the first well was
drilled in 1893, producing oil at a depth of 134 m. More systematic
geological exploration developed after 1923.

Among the largest oil fields are the Zamankuskoe, 120 km. west
of Groznyi on the North Osetian border, and the Karbulak-Achaluki,
70 km. west of Groznyi. Exploration in recent years has been successful in expanding reserves, mostly at considerable depths. The
average depth of wells is over 4,000 m., some reaching depths of
7,000 m. Drilling is often delayed and costs increased by the absorp-
tion of flushing media during the drilling of difficult rock strata.

However, because of the exploitation of deeper deposits, output
has been increased rapidly, rising from 9 million tons in 1965 to over
20 million in 1970, and is planned to reach 25 million tons in 1975.[34]

Krasnodar krai. The search for oil in the Kuban began in 1906, but
it was not until 1951 that any significant discoveries were made. The
largest oil field is the Anastasievsko-Troitskoe, which lies about 130
km. northwest of Krasnodar. Some new oil fields were found in the
1960s, when about ninety of the exploratory wells drilled were over
4,000 m. in depth, and the average depth for the 1970-75 period is
expected to be 4,710 m. Industrial diamonds are being used for deep
drilling. Unless further discoveries are made at depths less than
7,000 m. it will be difficult to maintain output at the 1972 level of 5.5
million tons.[35]

Stavropol krai. In 1965 there were nine oil fields in the Stavropol
krai, all discovered since 1945. The main producers are the Osek-
Suat and Velichaevka-Koledeznoe, situated about 100 km. east of
Prikumsk. From 1966 to 1970 intensive exploratory drilling was
undertaken, which led to the discovery of another 14 oil fields. In
1970 alone, 444,000 m. were drilled, with an average depth for exploratory wells of 3,400 m. Turbodrill speeds for development wells of
3,200 m. reached 1,415 m. per rig per month, while speeds of 2,301
m. per rig per month were recorded for new rotary drills.

From 1966 to 1970 production was up 50 percent from the previous ten years. About 4.6 million tons was produced in 1965 and
6.4 million in 1970, and 34 million is planned for the 1971-75 period,
which would suggest a yearly extraction rate of 7 million tons.[36]

Dagestan ASSR. Oil was extracted here in the nineteenth century using winches and buckets, and the first well was drilled in 1898. The Makhachkalinskoe oil field was discovered in 1943 on the Caspian coast. The eleven Dagestan oil fields had yielded a total of 7.5 million tons by 1965. This figure had been more than doubled by 1970, when annual output had risen to over 2 million tons.[37]

Petroleum in the North Caucasus is of better quality than in the Volga-Ural area, being low in sulphur with a paraffin base. Absolute output is being maintained because of the oil fields of the Checheno-Ingush ASSR, but the relative share in the USSR total is dropping.

Komi ASSR. Although signs of oil were recorded in the 17th century, systematic exploration did not begin until 1929. Several oil fields were discovered in the 1930s and 1940s between the Ukhta and Pechora rivers, and in 1959 the largest, Zapadnyi Tebuk, was found, 60 km. east of the town of Ukhta. Over 20 oil fields are now known in the Komi ASSR.

The "Komineft" union more than doubled output in the 1965-70 period, from 2.2 to 5.6 million tons. Since it had originally been planned to produce 8 to 10 million tons in 1970, there was some criticism for delays in bringing recent discoveries into production.[38]

Western Siberia

Western Siberia is without doubt the most important area in the future of the Soviet oil industry, and is therefore described here in greater detail.

The first prospecting was carried out between 1930 and 1933 and was continued on a more significant scale after the war (1947-50). In September 1953 some barges carrying rigs up the River Ob happened to be delayed near the village of Berezov, and it was decided to make a test boring near the river bank. This chance drilling was rewarded by a sudden blow-out of gas and water; the Berezovskoe gas field had been discovered.

It was rather more systematic explorations, however, that in 1960 revealed the first oil field, at the settlement of Shaim on the River Konda. Exploration has been steadily intensified; in 1972, for example, eight new oil fields were discovered in just the period from January to August, an average of one a month! The total number of oil fields discovered in Western Siberia by April 1973 was 113.

The quality of oils varies greatly, but they tend to have a low to medium sulphur content, under 2 percent.

There are important fields in both Tomsk and Novosibirsk oblasts, but by far the greatest number are in Tyumen oblast. It was claimed in 1972 that the reserves of the known oil fields formed but a small percentage of the potential resources of Western Siberia, which had been only partially explored. This was due to the vastness of the territory involved and certainly not to lack of effort.

From 1966 to 1970 about 3,200,000 m. were drilled in over 1,500 wells. Average drilling speed rose to 3,150 m. per rig per month. The leading brigades drilled over 60,000 m. in 1970, a rate of 5,000 m. per rig per month. The amount drilled each year is rising by an average of 180,000 m. and is expected to reach 2 million m. in 1975.

A large new research institute, "Giprotyumenneftegaz," which has been established in Tyumen, specializes in solving the problems faced by the oil industry in the north, such as the long severe winters and the vast roadless wastes of marshland and taiga. One of the most original ideas has been the use of hovercraft to transport drilling rigs. Experiments that have been in progress since 1966 have shown that the method is perfectly feasible; speeds up to 10 km. per hour have been attained with rigs of 160 tons weight, even over the bogs around Shaim. It has enabled a team of riggers to increase their construction capacity from 10 rigs in 1965 to 50 in 1970.

The greatest part of the known reserves $(A + B + C_1)$ is concentrated in just 23 large oil fields, which greatly simplifies extraction. There are three main centers for the oil industry in Western Siberia, the Shaim region in the Eastern Urals and the Surgut and Nizhnevartovsk regions around the middle reaches of the River Ob.

The exploitation of the Siberian oil fields began in 1964, and by 1972 twelve fields were already in production. There were two in the Shaim region: Trekhozernoe (1964) and Teterevo-Mortyminskoe (1966); five in the Surgut region: Ust-Balykskoe (1964), Zapadno-Surgutskoe (1965), Pravdinskoe (1968), Mamontovskoe (1970), and Solkinskoe (1972); and four in the Nizhnevartovsk region: Megionskoe (1964), Vatinskoe (1965), Sovetskoe (1966), and the biggest of them all, Samotlorskoe (1969). They were to be followed by the Fedorovskoe, near Surgut, in December 1972, and a further four oil fields were expected to be commissioned by 1975.

The rapid expansion of the oil industry in West Siberia is shown in Table 3.9. Output has increased at a much faster rate than in any of the other major producing areas in the USSR, mainly because Siberian oilmen have had the advantage of the most recent technological innovations. Of particular value is the utilization of local underground reservoirs of thermal mineral water with high flushing qualities for the maintenance of formation pressure; this has greatly simplified the problem of supplying water to the integrated pumping systems.

As may be seen in Table 3.9, the exploitation of new fields in the Nizhnevartovsk region (especially Samotlorskoe) in 1969 has brought a sudden increase in its share of Siberian oil production; this was 37 percent in 1970 and is likely to reach 68 percent in future years, to correspond more closely with its share in oil reserves.

TABLE 3.9

Exploitation of Oil Fields in Western Siberia, 1964-70

	1964	1967	1970	1964-70
Oil Output (in thousands of tons)				
Surgut	120	2,561	15,191	39,011
Nizhnevartovsk	73	945	11,588	20,273
Shaim	16	2,287	4,637	15,410
All three regions	209	5,793	31,416	74,694
Number of Development Wells				
Surgut	7	62	297	531
Nizhnevartovsk	5	34	159	276
Shaim	3	114	300	423
All three regions	15	210	756	1,230
Development Drillings (in thousands of metres)				
Surgut	3	215	506	1,545
Nizhnevartovsk	—	63	307	731
Shaim	9	133	158	791
All three regions	12	411	971	3,067

Sources: Efremov, E. P., et al., "Razvitie neftyanoi promyshlennosti Zapadnoi Sibiri v 1964-1970 gg.," Neftyanik no. 7, July 1971: 9.

Oil production in Western Siberia was expected to rise from 45 million tons in 1971, to 60 million in 1972, over 88 million in 1973, to 145 million in 1975, and 240 to 260 million in 1980. Pravda (29 March 1973) spoke of an annual output of 600 to 700 million tons, possible "in the near future." In 1975 it is planned to produce 78 million tons at the Samotlorskoe field alone. In July 1972 it was announced that 25 million tons of oil had been extracted at Samotlorskoe since exploitation began in 1969; output in 1973 was to be 38 million, and an annual production rate of 100 million is considered possible in the near future. In March 1973 it was reported that a total of over 200 million tons had been produced in Western Siberia.

The main difficulty experienced in the 1960s was finding an outlet for the oil that was being extracted in such large quantities. There were not enough refineries or petrochemical plants in Siberia itself, and transport facilities were inadequate. There were delays in the construction of the pipeline from Ust-Balyk to Omsk. Rivers were the main transport arteries, but could not be used when frozen during the long winters, and there was also a shortage of shallow-draft tankers.

The transport situation, however, is steadily improving with the completion of more oil pipelines and the Tyumen-Tobolsk-Surgut railway, as well as many new roads and the river ports at Tobolsk and Surgut. Refinery capacity is being increased to absorb oil production more effectively, and the Surgut power station will consume large quantities of incidental oil-well gas. When the Tobolsk petrochemical complex is completed, the utilization of incidental gas should reach 85 percent, but the average over the 1971-75 period will be below 40 percent. Before 1970 almost all incidental gas in Western Siberia was vented and flared, wasting a valuable fuel and raw material for the chemical industry.

There have been many other problems to overcome in the course of developing the Siberian oil industry. The harsh climate has caused much delay, which when aggravated by poor cooperation between various sectors of the Soviet economy has sometimes made it almost impossible to integrate the various stages in operations. Wells have been prepared for production before there were pipelines ready to transport the oil, and possible expansion in the output of existing wells has often been limited because pipelines were working at full capacity and were unable to cope with additional oil. Pipeline specialists developed a method of increasing the yearly capacity of the Ust-Balyk to Omsk pipeline by 2 million tons, but this depended on an additional supply of electric power, which was not available until the Aremzyan sub-station and the transmission line from Vagai to Novopetrovo were completed. The cement supplied for well construction has often been of poor quality, and many major repairs have been required.

In 1970 some 44 million cu.m. of water were pumped into formations to maintain pressure, and one of the main problems facing Siberian oilmen is raising this volume to over 200 million cu.m. by 1975. Trouble has been experienced at the Pravdinskoe field, where formation pressure has dropped.

Soviet experts are nonetheless optimistic about the future of the oil industry in Western Siberia, and their views are indeed solidly based on the impressive achievements of recent years. From 1964 to 1970 Western Siberia received 20 percent of the total capital investment in the oil industry. It is estimated that the greatly increased capacity for production attained over this period cost 32 percent less per ton than the average for the industry as a whole, and it is expected that returns for investment will continue to improve in this new oil center, the "third Baku."[39]

Eastern Siberia

Prospecting in Eastern Siberia began in 1939 but was suspended until 1948 because of the war. Three oil fields have been found: the Markovskoe, on the Lena river, southeast of Kirensk, and the Atovskoe and Yuzhno-Raduiskoe, between the Angara river and Lake Baikal, north of Irkutsk. The oil from these fields is of very high quality, and further exploration is being undertaken to establish a new base for the Soviet oil industry.[40]

Sakhalin

Exploration began in the island of Sakhalin in 1922 and was concentrated mainly in the northeast, where some thirty oil fields have been discovered. Sulphur content is low. By 1965, when 2.4 million tons were extracted, a total of 24 million tons had been produced since the first field was brought into exploitation in 1923. Yearly output had not been significantly increased by the beginning of the 1970s.[41]

Some important production indices that demonstrate the leading role of the RSFSR in Soviet oil are given in Table 3.10. Output by oil association is shown in Table 3.11, and distribution of oil production by regions of the USSR is summarized in Table 3.12.

Azerbaidzhan

In 1972 the second largest oil producer among the union republics was still Azerbaidzhan, although it had been surpassed by five separate administrative areas within the RSFSR: Tatar ASSR, Bashkir ASSR, Kuibyshev oblast, Chechen-Ingush ASSR, and West Siberia. As mentioned above, oil was being extracted in Azerbaidzhan even before the 17th century; and before the discovery of oil in Texas in 1901, the Baku oil fields were the most important producers in the world.

On 28 March 1971 the Azerbaidzhan oilmen celebrated the extraction of a total of 1,000 million tons since exploitation began. Production has been mainly from the fields concentrated in the Apsheron peninsula, around Baku (80 percent); from the Apsheron archipelago (10 percent); and from the Kura river fields (3 percent).

TABLE 3.10

Oil Extraction in the RSFSR, 1971

Area	Output (in millions of tons)	Cost per Ton (as a percentage of USSR average)	Labor Productivity (in tons per man for one year)
USSR	371.8	100	4,365
RSFSR	300.7	85.4	5,148
Tatar ASSR	101.0	65.8	8,517
Kuibyshev oblast	35.4	68.6	6,233
Perm oblast	16.9	73.4	5,779
Western Siberia	44.7	76.4	18,472
Chechen-Ingush ASSR	21.6	66.0	7,077

Source: Brenner, M. M., "Razvitie neftyanoi promyshlennosti SSSR i ee syrevoi bazy," Neftyanoe khozyaistvo no. 12, December 1972: 12.

There are about 60 oil fields, among the most important of which are the Lobatinskoe, 10 km. southwest of Baku; Karadag, 43 km. southwest of Baku; Neftyanye Kamni, in the Caspian Sea, 120 to 140 km. southeast of the Apsheron peninsula; and Sangachaly-Duvannyi, in the Caspian, 41 km. southwest of Baku.[42]

In recent years exploration has been most successful in the offshore areas, particularly around Duvannyi and Bulla islands, and it is hoped that further reserves will be discovered deep under the Caspian. Wells have been sunk to depths of almost 6,000 m. in difficult structures, but drilling plans for the 1966-70 period were not fulfilled. Exploration on land has been disappointing, and future development will be concentrated on the area under the sea.

Output since the war has remained below the 1940 level; although yearly production climbed slowly to 21.7 million in 1966 it has since dropped gradually to under 19 million in 1972, when over two-thirds was from the new fields.

It is possible that some oil will be extracted from mines at Balakhony and from oil-saturated sands by open-pit mining at Kirmaku, without the cost exceeding the average in the older mainland fields.

TABLE 3.11

Oil Output, by Association, 1970-73
(in millions of tons)

Association	1970	1972	1973 (planned)
Tatneft	100.4	102.1	103.2
Glavtyumenneftegaz	31.4	62.7	86.5
Bashneft	40.7	40.1	40.6
Kuibyshevneft	34.9	35.5	35.9
Grozneft	20.3	19.9	16.5
Mangyshlakneft	10.4	15.1	17.1
Permneft	16.1	17.8	19.2
Turkmenneft	14.4	15.8	16.0
Ukrneft	13.5	13.9	13.3
Kaspmorneft	12.9	11.8	11.9
Orenburgneft	7.4	9.4	10.5
Nizhnevolzhskneft	7.0	7.7	7.6
Stavropolneftegaz	6.4	6.9	7.0
Komineft	5.6	6.3	6.8
Azneft	7.3	6.6	6.3
Belorusneft	4.2	5.9	7.0
Krasnodarneftegaz	5.3	5.5	5.6
Embaneft	2.7	3.0	3.2
Sakhalinneft	2.5	2.4	2.4
Dagneft	2.2	2.0	1.8
Udmurtneft	0.5	1.3	2.3

Source: Ekonomicheskaya gazeta no. 5, January 1973: 2.

Other interesting developments are in offshore exploitation, where output has been rising in recent years, unlike the general trend on the Azerbaidzhan mainland. In 1971 about 35 percent of oil production was from the Neftyanye Kamni fields, at a third of the average cost on the mainland. By injecting water or gas into the formations oilmen have kept oil flowing from wells even twenty years after they first gave oil. The first rigs were constructed on the hulls of old ships that had been sunk in position. Several lives were lost before the long sea gantries, which are now famous throughout the USSR, were properly established in the early 1950s. There are now over 300 km. of raised platforms and roads twisting over the surface of a sea known

TABLE 3.12

Regional Distribution of Oil Production, 1965-75
(in millions of tons)

Region	1965	1970	1972	1975 (planned)
USSR	242.9	352.6	394.0	496
RSFSR	199.9	284.3	321.0	400
Of which Volga-Ural	173.6	207.1	213.9	210 to 225*
Tatar ASSR	79.7	100.4	102.1	100
Bashkir ASSR	40.7	40.7	40.1	40
Kuibyshev oblast	33.4	34.9	35.5	35
Perm oblast	99.7	16.1	17.8	27
Orenburg oblast	2.6	7.4	9.4	12
Volgograd oblast	6.2	6.0*⎫	7.7	NA
Saratov oblast	1.3	1.0*⎭		NA
Udmurt ASSR	—	0.5	1.3	6
North Caucasus	20.8	34.2	33.3	35 to 45*
Chechen-Ingush ASSR	9.0	20.3	19.9	25
Stavropol krai	4.6	6.4	6.9	7*
Krasnodar krai	6.2	5.3	5.5	NA
Dagestan ASSR	1.0	2.2	2.0	NA
Western Siberia	1.0	31.4	62.7	125
Komi ASSR	2.2	5.6	6.3	10
Sakhalin	2.4	2.5	2.4	NA
Azerbaidzhan SSR	21.5	20.2	18.4	19*
Turkmen SSR	9.6	14.5	15.9	22
Kazakh SSR	2.0	13.2	18.1	30
Of which Mangyshlak	0.3	10.4	15.1	27
Ukraine SSR	7.6	13.9	14.5	15
Belorussian SSR	—	4.2	5.8	8
Uzbek SSR	1.8	1.8	1.4	1

NA: Not available.

*Approximate figure, estimated by the author from incomplete data.

Sources: Narodnoe khozyaistvo SSSR v 1969 godu, Moscow 1970; SSSR i soyuznye respubliki v 1970 godu, Moscow 1971; Direktivy XXIV sezda KPSS po pyatiletnemu planu razvitiya narodnogo khozyaistva SSSR na 1971-1975 gody, Moscow 1971; Ekonomicheskaya gazeta no. 6, January 1973: 2; Narodnoe khozyaistvo SSSR 1922-1972, Moscow 1972; Further data from the relevant section of the text, and from Table 3.7.

for its sudden storms and strong winds. Living conditions are reasonable, with houses, restaurants, libraries, museums, and even hothouses to raise flowers, which are later transferred to open-air beds.

Production in Azerbaidzhan is expected to continue to fall, since the new areas will not compensate for the exhaustion of the old fields.[43]

Turkmenia

Oil has been extracted in the Cheleken peninsula since the 19th century, but the greatest development has taken place since the last war. There are four important oil fields in Turkmenia: Nebit-Dag, Kum-Dag, Cheleken, and Kotur-Tepe, and a few other large oil-gas fields also in the west, near the Caspian Sea.

From 1873 through 1965 about 86 million tons of oil were produced in Turkmenia, but in the following five years some 64 million tons were extracted, 14.5 million in 1970 alone.

The average depth of exploratory wells has increased from 2,500 m. in 1965 to over 3,000 m. in 1970, and some new deposits of oil have been discovered. The most interesting expansion, however, is taking place off the eastern shore of the Caspian, where complexes similar to those of Neftyanye Kamni are being constructed. An oil port is taking shape on the Cheleken peninsula, which is to be connected by rail to the main Central Asian railway.

It is expected that 3 million out of the 22 million tons Turkmenia is planned to produce in 1975 will be from offshore operations in the Caspian.[44]

Kazakhstan

Oil has been extracted in the Emba area around Dossor since 1909, but the rate of growth was so slow that by 1965 total yearly production was little more than 2 million tons from over twenty fields. In recent years, however, output has increased rapidly with the discovery of new fields, and Kazakhstan is expected to contribute significantly to future expansion in the Soviet oil industry.

The Koschagyl field, which has produced over 4 million tons since 1935, is situated about 100 km. inland from the northeast shore of the Caspian. The Prorva field on the coast and the Kenkiyak field about 125 km. southeast of Aktyubinsk are also among the important fields in the northwest of Kazakhstan.

Since 1961 several fields have been discovered in the Mangyshlak peninsula, where the greatest increase in output is now taking place. Those at Zhetybai and Uzen are the largest, with about 13 million tons produced in 1973. The "Mangyshlakneft" union was able to increase its output from .3 million tons in 1965 to 10.4 million in 1970, and it is planned to produce 26.5 million in 1975. It is thought that annual output could reach 100 million tons in Mangyshlak.

Yearly production in Kazakhstan as a whole rose from 3.1 million tons in 1966 to 13.2 million in 1970, and this spectacular progress is expected to continue throughout the 1970s, reaching 30 million tons in 1975.

At the Uzen field pressure is maintained by injecting hot water into the formations, and in 1970 an experimental system was instigated to pump up to 15,000 cu.m. of sea water at 95° C into the oil-bearing layers. The most successful method in wide use, however, is extraction by gas pressure.

A new oil town called Novyi Uzen has been built in the arid desert near the oil field. It now has a population of over 40,000. Fresh water is supplied to the new oil area by a pipeline stretching 250 km. from the nuclear power station and desalination plant at Shevchenko. Alternative methods of supplying fresh water, such as by pipeline from the Amu-Darya river or even from the Volga, are being considered.[45]

Uzbekistan

Compared with Turkmenia, the other republics of Central Asia are not very significant producers of oil. There are some thirteen oil fields and over twenty oil and gas fields in Uzbekistan, mainly in the Fergana valley, but the total yearly output of oil in 1970 was only 1.8 million tons. Some new fields were discovered in 1967 and 1968, but the oil fields in general are small and are approaching exhaustion.[46]

Oil has been produced in Tadzhikistan since 1909, but total output since then has been less than 2 million tons, and yearly output by 1972 was still only 198,000 tons. Production in Kirgizia is also comparatively insignificant and has actually fallen from 464,000 tons in 1960 to under 300,000 tons in 1972.[47]

Ukraine

Because of the concentration of industry and population in the European part of the USSR, oil produced locally is of particular economic value. Output in the Ukraine has grown steadily since 1960 and now makes a significant contribution to the Soviet fuel economy.

Drilling for oil began in 1875, and oil was first extracted in commercial quantities at Borislav in 1909, but it is in more recent years that the most important discoveries have been made.

There are three main oil-bearing areas: in the west, towards the Carpathian mountains; in the east, between Kharkov, Kiev, and Dnepropetrovsk; and in the south, around Kerch in the Crimea. Until recently production was from the small fields in the west, but after

1965 several new fields were discovered in the east, the largest being the Gnedintsevskoe and the Lelyakovskoe. In 1970 these two fields produced 7.3 million tons, or 54 percent of the total for the Ukraine.

Output has risen steadily from 2.2 million tons in 1960 to 14.5 million in 1972, and it is planned to extract over 15 million tons in 1975. Over 80 percent is from the fields in the east of the Ukraine.

Drilling is becoming more complicated, since many wells are now being drilled to depths between 4,000 and 7,000 m., but exploration is continuing because about 44 percent of the territory of the Ukraine is considered potentially oil-bearing.[48]

Belorussia

Prospecting began as recently as 1961, and oil was being extracted by 1965, when 39,000 tons were produced. Output increased rapidly to 1.7 million tons in 1968, 2.7 million in 1969, and 5.8 million in 1972, with 7 million planned for 1973. One of the first fields to produce oil was the Rechitskoe, near the town of Gomel.[49]

Remaining Regions of the USSR

Even more recent developments have taken place in Lithuania and in the neighbouring Kaliningrad oblast of the RSFSR, where the Ushakovskoe field was prepared for industrial exploitation in 1972, yielding 300 tons a day by July. A further four fields are to be brought into production by 1975. According to a Moscow radio broadcast, in May 1972 known oil reserves in Kaliningrad oblast exceeded 11 million tons.[50]

Oil has been found in Moldavia and Georgia, but production is too low to be of more than local significance.

Technological Progress

Methods of oil production have been greatly improved in recent years, and this has contributed greatly to the rapid growth rate in the Soviet oil industry.

The most important single factor has been the development of pressure maintenance projects since 1945. To cite but one example, the water-injection system at the Romashkinskoe field in the Tatar ASSR has made it possible to extract some 175 million tons more oil than would have been possible by prewar methods, and the saving in capital investment because fewer wells were needed has been calculated at 2,000 million rubles.

Out of 449 oil fields in exploitation in 1971, pressure was maintained in 168 of them by water-injection systems. In the 1966-70 period these systems increased output by over 660 million tons.

Some wells are exploited by pumping water heated to 350°C at 100 to 150 atmospheres of pressure into the layers. By using such new technology as this it is hoped to raise the proportion of oil recovered to between 60 and 70 percent.

Some indication of trends in oil production methods is given in Table 3.13. The most economical form of extraction is by flowing well, but the number of wells producing by this method has been falling since it reached 74 percent in 1961 because many of the largest among the old oil fields are approaching exhaustion. With progress in technology, labor productivity has been greatly improved.

Several new systems have been developed to make the automation of the whole process of extraction possible. This is of particular importance in Western Siberia, which suffers from rapid labor turnover because of its hostile climate, even though bonuses of up to 70 percent are paid on top of the normal basic wage.

Automatic equipment that regulates such factors as oil flow and water pressure has been in use for several years, but in 1972 a new system devised by the research institutes of Baku and Groznyi was installed in several oil fields. Known from its Russian initials as "PAT," it can automatically monitor up to forty scattered production units by remote control.

By 1975 as many as 17 PAT systems and 300 of the less complex "Sputnik" systems will have been installed in the Tyumen fields alone. A computer center to simplify long-term planning and prospecting operations will also be established in Tyumen.

Further advances are being made in the extraction of oil from the sea bed, especially in the Caspian fields off Azerbaidzhan. In 1970 over 13 thousand tons of oil were extracted from the sea bed, at a cost per ton that was below the average for the land fields in Azerbaidzhan, Tataria, and Kuibyshev oblast. Floating drilling rigs built in Baku have successfully drilled test bores of up to 1800 m. when positioned in heavy seas far from the shore at depths up to 300 m. In August 1972 sea wells were drilled to a depth of 5300 m. in the Bulla area, the deepest ever drilled in the Caspian.

Methods of extracting oil simultaneously from several layers through a single well are being improved. Such systems bring considerable savings in steel, cement, and working time.[51]

Some hermetically-sealed systems for the combined extraction of oil and incidental gas have been installed. They incorporate all the processes as far as the distribution point and have greatly increased the utilization of incidental gas, which formerly needed a separate system. In 1970 only 62.3 percent of this valuable raw

TABLE 3.13

Oil Production Methods in USSR, 1950-70
(selected indicators)

	1950	1960	1965	1970
Method of production (per unit of total)				
Flowing Wells	32.5	73.7	64.4	55.4
Pumped Wells	44.7	23.5	33.5	42.7
Compressed air/gas	21.1	2.3	1.8	1.8
Others	1.7	0.5	0.3	0.1
Percentage extracted by pressure maintenance	23.0	65.0	67.0	70.0
Water injected (in millions of cu.m.)	8.9	189.5	329.1	550.0
Air and gas injected (in millions of cu.m.)	343.7	563.1	469.7	NA
Output per well per month (in tons)	170.0	371.0	510.0	595.0*
Wells in first year of exploitation	510.0	1047.0	1062.0	1380.0*
Old wells	153.0	349.0	499.0	575.0*
Tons of oil produced per worker	710.0	2087.0	2379.0	3284.0

NA: Not available.

*Estimated from 1966-69 figures.

Sources: Melnikov, N. V., Mineralnoe toplivo, Moscow 1971: 49; Narodnoe khozyaistvo SSSR v 1969 godu, Moscow 1970: 198; Neftyanik no. 2, February 1971: 2.

material was gathered, and utilization was particularly bad in the Tyumen, Orenburg, Mangyshlak, and Turkmen oil unions.

In the 1971-75 period it is planned to introduce automatic systems in over 120 oil-extraction enterprises, with some 30,000 wells and a yearly output of 200 million tons. Progress in this respect has not been as rapid as had been hoped. The head of the administration for technological development in the oil and gas industry, Yu. Zaitsev, admitted in 1971 that "the basic problem of the complex automation of oil-extracting enterprises is not yet being solved in a completely satisfactory manner." He blamed the leaders of the Perm, Komi, Kuibyshev, Turkmen, Tatar, Emba, and Sakhalin unions for not producing the necessary information and plans and for failing to apportion sufficient capital investment for the project.

Although improvements in technology have been considerable, many criticisms are published in the Soviet press attacking research institutes and the ministries manufacturing equipment for delays and for poor quality products. It is occasionally admitted that American oil-extracting technology is ahead of the Soviet equivalent in many cases, and it is doubtless hoped that the USSR will benefit from the improved trade conditions of the 1970s by importing equipment from the United States. By the early 1970s this trend had already become evident. In 1970 the USSR spent 20.7 million rubles on importing equipment for what was termed "geological exploration, engineering geology, and the extraction of oil and natural gas." Rumania supplied 12.6 million rubles worth of equipment, Britain 5.3 million, and France 2.1; the contribution of the United States was negligible. By 1972 the total expenditure on importing this category of equipment was 36.7 million rubles, distributed as follows: Rumania 20.7, Britain 8.5 million, the United States 3.8 million, and France less than one million.[52]

Transportation

In the early years of the Soviet oil industry refineries were concentrated in the areas of extraction, around Baku and Groznyi. With the rapid growth in demand for oil products, however, many refineries were built in the centers of consumption, and the efficient transporting of crude oil became of vital importance to the development of the economy.

The optimum yearly capacity for the Soviet refineries at present being planned is put at 12 to 15 million tons, but refineries of 18 to 24 million tons capacity per annum are being considered. The largest oil refinery centers have been established in the Ural, Volga, Central, Siberian, and Southern economic regions. In 1973 the total capacity of the refineries in Comecon countries was put at 360 million tons annually, of which about 350 million tons was in the USSR.[53]

The railway network and the Volga-Caspian shipping routes no longer account for most of the oil transported in the Soviet Union. In 1940 about 45 percent of oil and oil product transporting was by rail, 44 percent by ship, and only 11 percent by pipeline. By the beginning of the 1970s the share of rail transport had dropped to about 40 percent, and transport by river and sea had fallen considerably to under 10 percent, while pipelines accounted for over 50 percent of all movement of oil and oil products. Transport by road is comparatively insignificant.[54]

The growth in the volume of crude oil and oil products being transported by these methods in the USSR is shown in Table 3.14. Although pipeline transportation of oil is generally considered the most efficient method when such factors such as steel utilization, energy consumption, and labor productivity are compared, pipeline construction in the USSR has only recently begun to keep pace with the increase in oil output. Some indication of the importance of rail transport is given by the fact that in 1969 oil and oil products accounted for 342,800 million ton-kilometers, and the average length of haul per ton was 1,204 km., both indicators being significantly higher than those for pipelines.[55]

It is expected, however, that by 1980 pipelines will have taken over the transportation of oil and oil products almost completely. The development of the trunk pipeline system in the USSR is shown in Table 3.15 and Map 3.2. The accelerated rate of construction that began after 1950 is striking: 5,000 km. were added to the network by the end of 1955, an additional 7,000 km. by 1960, and as much as 11,000 km. by 1965. The rate of construction dropped slightly in the following five years, but over 22,000 km. of new trunk lines are planned for the 1971-75 period. The actual construction achieved in 1971 and 1972, however, made this target seem improbable.

As might be expected in view of the large disparity in oil consumption, the United States has a much more extensive pipeline network, which is continuing to grow at a considerable rate. (See Table 3.16.) It should be noted, however, that the USSR has a higher percentage of large-diameter pipeline than the United States.

Until 1950 the maximum pipeline diameter was only 350 mm. and the capacity of pumps was below 135 cu.m. per hour. By 1967 small-diameter pipeline formed only 4 percent of the network; 27 percent was of 529 mm. diameter pipe; 27 percent of 720 mm.; 18 percent of 820 mm.; and 19 percent of 1,020 mm. The share of 1,020 mm. and 1,220 mm. diameter pipeline has been rising steadily in recent years. There are pumps now in use that have capacities of 7,000 to 10,000 cu.m. per hour.

The cost of transporting oil by pipeline is now on average less than a third of the equivalent cost by rail.

110

TABLE 3.14

Transportation of Oil and Oil Products, 1940–71
(in millions of tons)

Method of Transportation	1940	1950	1960	1965	1968	1969	1970	1971
Rail	29.5	43.2	151.0	222.2	275.9	284.7	302.8	322.8
Sea	19.6	15.8	32.5	53.5	70.1	70.5	75.1	79.8
River	9.6	11.8	18.4	25.0	29.2	30.3	33.5	35.2
Pipeline	7.9	15.3	129.9	225.7	301.3	324.0	339.9	352.5

Sources: Narodnoe khozyaistvo SSSR v 1969 godu, Moscow 1970: 445, 455, 458, 468; Narodnoe khozyaistvo SSSR 1922–1972, Moscow 1972: 297, 301, 303, 306.

TABLE 3.15

Construction and Throughput of Oil Pipelines in
USSR, 1913-75

Year	Length at End of Year (in thousands of km.)	Throughput of Oil and Oil Products (in millions of tons)	Turnovers (in thousands of millions of ton-km.)
1913	1.1	0.4	0.3
1921	1.1	0.3	0.1
1928	1.6	1.1	0.7
1932	2.9	4.8	2.9
1937	3.9	7.5	3.6
1940	4.1	7.9	3.8
1945	4.4	5.6	2.7
1946	4.4	6.0	2.9
1950	5.4	15.3	4.9
1955	10.4	51.7	14.7
1956	11.6	65.3	20.5
1957	13.2	80.9	26.6
1958	14.4	94.9	33.8
1959	16.7	111.3	41.6
1960	17.3	129.9	51.2
1961	20.5	144.0	60.0
1962	21.7	165.1	74.5
1963	23.9	185.5	90.9
1964	26.9	213.0	112.1
1965	28.2	225.7	146.7
1966	29.5	247.7	165.0
1967	32.4	273.3	183.4
1968	34.1	301.3	216.0
1969	36.9	324.0	244.6
1970	37.4	339.9	281.7
1971	42.9	388.5	375.9
1972	45.1	NA	NA
1973	49.3	NA	NA
1974*	54.5	NA	NA
1975*	60.0	430.0 (crude oil)	

*Planned.

NA: Not available.

Sources: Narodnoe khozyaistvo SSSR v 1969 godu, Moscow 1970: 468;
Neftyanik no. 1, January 1972: 7; Neftyanik no. 2, February 1972: 2; Narodnoe
khozyaistvo SSSR 1922-72, Moscow 1972: 306.

112

MAP 3.2

Major Oil and Oil Product Pipelines

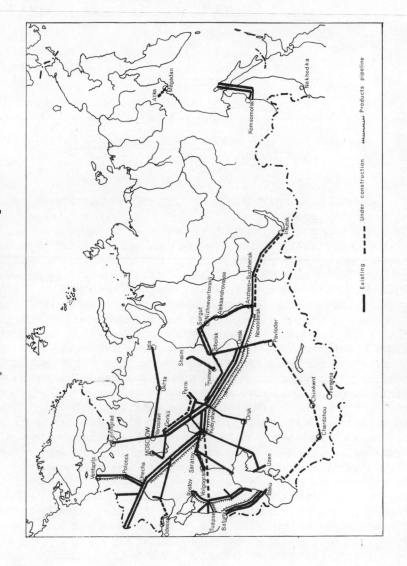

Source: Updated from <u>Sovetskii Soyuz—Obshchii obzor</u>, Moscow 1972: 690-91.

TABLE 3.16

Comparison of Oil Pipeline Networks in
USSR and United States
(trunk lines in thousands of km.)

Year	USSR	United States
1950	5.4	129.6
1955	10.4	144.0
1960	17.3	164.8
1965	28.2	184.0
1969	36.8	200.0

Sources: Table 3.15 and Statistical Abstract of the United States 1971, Washington 1971: 550.

In 1970 about 8,000 km. of the total oil pipeline network was made up of oil product lines.[56]

The construction of pipelines before the 1960s has been described above; perhaps the most significant addition to the pipeline network in the 1960s was the Druzhba (Friendship) pipeline, which supplies Soviet oil to Czechoslovakia, East Germany, Hungary, and Poland. Agreements were signed by the governments of the Comecon countries in December 1959, and construction started in 1960. An unfortunate aspect of this event was the decision taken in 1962 by NATO countries not to sell large-diameter pipe to Eastern Europe in order to delay construction of Druzhba, which they feared would increase the mobility of the Warsaw Pact armies. West Germany was particularly affected by this embargo because it had previously been supplying a large part of the Soviet requirements. Completion of the system was actually retarded by about a year, but since the embargo was not immediately applied, and since Sweden was not in any case affected, over a million tons of large-diameter pipe was received from the West. Moreover, pipe-rolling technology in the USSR was given a lasting impetus, and large-diameter pipe can now be produced at plants in Novomoskovsk, Zhdanov and Khartsyzsk in the Ukraine, and in Chelyabinsk in the Urals; the technology of manufacturing pipes of as much as 2,520 mm. diameter has been mastered. The embargo was revoked in 1966, and imports of pipe have been increasing over recent years. (See Chapter 8.)

Pumping of crude oil from the Urals to Czechoslovakia began as early as February 1962, and by 1963 oil was also reaching Poland,

East Germany, and Hungary. The line runs west from Kuibyshev through Penza, Michurinsk, Orel, and Bryansk and branches to Polotsk and Ventspils at Unecha. Further west at Mozyr it divides, with one branch going through Brest to Plock in Poland and on to Schwedt and Leuna in East Germany, while the other runs through Lvov and Uzhgorod to Szazhalombatta, just south of Budapest, and to Bratislava and Zaluzi-Most in Czechoslovakia.

Out of a total length of 5,327 km., 3,688 km. are in the USSR, including the Unecha-Polotsk-Ventspils branch; 826 km. in Czechoslovakia; 656 km. in Poland; 130 km. in Hungary; and 27 km. as far as Schwedt in East Germany.

The pipeline, which varies in diameter from 426 mm. to 1,020 mm., crosses terrain as high as 1,100 m. above sea level and traverses such rivers as the Volga, Dnepr, Vistula, Oder, and Danube. In 1968 it was extended 350 km. from Schwedt to Leuna, and in 1969 from Schwedt 250 km. to Rostok. In 1972 a second pipeline was being constructed parallel to Druzhba.

Pumping units with a capacity of 7,000 cu.m. per hour were first used in the Druzhba system, which carries about 19 million tons of oil a year. When the second line is completed the two will have a combined capacity of 50 million tons a year. By the beginning of 1973 some 210 million tons had been supplied by the USSR to Czechoslovakia, East Germany, Hungary, and Poland via the Druzhba system.

An oil pipeline of 720 mm. diameter running from Almetevsk through Gorkii and Ryazan to Moscow was begun in 1958 and completed in 1962. As mentioned above, the Trans-Siberian crude oil pipeline (Ufa-Omsk-Novosibinsk-Krasnoyarsk-Irkutsk) went on flow in 1964, moving oil at one-quarter of the cost of rail transport. In 1965 the first oil pipeline in Western Siberia was completed; the whole operation, from planning to building the 410 km. of 529 mm. diameter pipe from Shaim to Tyumen, took only 22 months in spite of the difficult conditions. In the same year the pipeline from Uzen through Zhetybai to Shevchenko went on flow.

In 1967 the second crude oil pipeline in Western Siberia (Ust-Balyk through Tobolsk and Tara to Omsk) was completed, using pipe of 1,020 mm. diameter and covering almost 1,000 km. By 1969 a pipeline from Aleksandrovskoe through Nizhnevartovsk to Ust-Balyk had also been built. In 1972 the underground pipeline from Aleksandrovskoe to Andzhero-Sudzhensk was completed over a distance of 818 km., the first oil pipeline of 1,220 mm. diameter pipe. It was built through deep snow, marshes, and forests and across the Ob, Tom, Vasyugan, and Parabel rivers. It was sometimes necessary to work in temperatures below -50° C.

The importance of this pipeline is very great indeed, since the Siberian oil fields are expected to contribute as much as 125 million tons in 1975 and double that amount in 1980. The 818 km. of pipeline required 328,000 tons of 1,220 mm. diameter steel pipes and is to have ten pumping stations with pumps of 10,000 cu.m. per hour capacity, powered by 6,300 kw. electric motors. In 1972 only two pumping stations were in commission, the head station at Aleksandrovskoe, which has storage tanks for up to 4.6 million tons of oil, and a second station in the Parabel region.

The completion of the Siberian oil pipeline ring means that oil from the Aleksandrovsk region no longer has to be pumped eastward through Omsk and can save some 1500 km. on its route to the Angarsk and Khabarovsk refineries, thus cutting transport costs by 50 percent. By using pipes of 1,220 mm. rather than 1,020 mm. diameter capital investment reduced about 20 percent and throughput is increased by about 50 percent. Automation of control will keep service personnel to the minimum. Capital investment in the pipeline should be recovered in only two years.

The Trans-Siberian underground trunk line from Anzhero-Sudzhensk to Irkutsk was no longer adequate to meet requirements; some 60 million tons of crude oil had to be transported in 1972 and as much as 125 million tons will have to be transported in 1975. A second line to Irkutsk, which was to be extended right to the Far East port of Nakhodka, was therefore begun in 1972. The importance of this development is obvious, particularly in view of growing Japanese and American interest in Siberian oil.

In 1972 a pipeline from Ukhta to Yaroslavl was built over a distance of 1,135 km., and a 1,844 km. pipeline from Ust-Balyk through Kurgan and Ufa to Almetevsk was begun, being completed in record time by April 1973. It has 24 pumping stations.

Another important feat of technology is the construction of a pipeline from Uzen through Gurev to Kuibyshev. As Mangyshlak high-paraffin crude congeals at temperatures around 30° C, this 1,020 mm. diameter pipeline has had to be constructed with heating installations in order to maintain flow.

In January 1972 the oil product pipeline from Polotsk to the Baltic port of Ventspils was completed parallel to the existing oil pipeline, allowing easier delivery of diesel fuel.

In 1972 some important projects were initiated. Particularly impressive was the plan to build a pipeline from Omsk through Pavlodar, Karaganda, Temirtau, and Dzhezkazgan to Chimkent, from where it will continue across Uzbekistan to the refinery at Chardzhou in Turkmenia.

Construction has begun on a pipeline of 1,400 km. to run from Kuibyshev through Tikhoretsk to the Black Sea port of Novorossiisk. It is to be completed in 1975 and will deliver oil from the Volga-Ural, Western Siberia, and Mangyshlak fields.

A pipeline was begun from Michurinsk to Kremenchug that was to be put into operation in 1973 and then continued to Kherson and Odessa. The Monastyrishche-Priluki line was extended to Kherson in 1972.

The pipeline from Ukhta to Inta was completed by summer 1973, and the second line from Unecha to Polotsk was delivering oil to the Novopolotsk refinery early in the same year.[57]

These and other developments are shown in Map 3.2.

The oil pipeline network in the USSR is thus expanding at a rate that more than keeps pace with the rapid growth of the oil extraction industry, and it is expected to gradually relieve the railway system of almost all long-distance haulage.

The export and consumption of oil in industry is discussed in Chapter 8.

NOTES

1. Mirzoev, Kh. I., Iz istorii poiskov nefti v Azerbaidzhane (From the history of oil prospecting in Azerbaidzhan), Baku 1970, 12-83; Bakirov, A. A., Ryabukhin, G. E., Neftegazonosnye provintsii i oblasti SSSR (Oil and gas-bearing provinces of the USSR), Moscow 1969.

2. Ponomarev, B. N., et al, Istoriya SSSR (History of the USSR), 12 vols., Moscow 1967, Vol 7, 611-16; Chamberlin, W. H., The Russian Revolution, 2 vols., New York 1965, Vol 2, 410-15.

3. Mirzoev, op. cit., 101-102.

4. Budkov, A., Budkov, L., "Neftyanaya promyshlennost v resheniyakh Kommunisticheskoi partii i Sovetskogo pravitelstva" (Oil industry in the decisions of the Communist party and Soviet government), Neftyanik no. 3, March 1971: 1-3; Bakirov, op. cit.: 5.

5. Kostrin, K. V., "70 let otechestvennomu truboprovodostroeniyu" (70 years of Soviet pipeline construction), Stroitelstve truboprovodov no. 1, January 1970: 78; Kostrin, K. V., "I. P. Ilimov—initsiator stroitelstva magistralnykh nefteprovodov v Rossii" (I. P. Ilimov—first builder of main oil pipelines in Russia), Neftyanik no. 3, March 1970: 34-35.

6. Budkov, op. cit., 3-4; Melnikov, N. V., Mineralnoe toplivo, Moscow 1971: 37.

7. Pospelov, P.N., et al., Velikaya Otechestvennaya voina Sovetskogo Soyuza 1941-45 (Great Patriotic War 1941-45), Moscow 1967: 181, 278.

8. Zakharov, M. V. et al., 50 let vooruzhennykh sil SSSR (50 years of the Soviet armed forces), Moscow 1968: 457; Worth, A., Russia at War, London 1965: 568.

9. Ekonomicheskaya gazeta no. 27, July 1972: 1; Budkov, op. cit.: 4.

10. Melnikov, op. cit.: 37.

11. Kostrin, K. V., "70 let otechestvennomu truboprovodos-troeniyu," Stroitelstvo truboprovodov no. 1, January 1970: 8, 9; Neftyanik no. 2, February 1972: 2.

12. Melnikov, op. cit.: 40-47; Neftyanik no. 2, February 1972: 20, 21.

13. Energeticheskie resursy SSSR—Toplivno-energeticheskie resursy (Energy resources of the USSR—fuel resources), Moscow 1968: 276-80; Pravda, 20 April 1973: 2.

14. Ibid.: 287-91; UN Statistical Year Book 1970, New York 1971: 212-13.

15. Melnikov, op. cit.: 35-36.

16. Galonskii, P. B., "Bolshie perspektivy, vazhnye zadachi" (Great prospects, important problems), Neftyanik no. 2, February 1972: 2.

17. Rozen, V., "Poiski i dobycha morskoi nefti" (Prospecting and extracting sea oil), Neftyanik no. 10, October 1971: 33-34.

18. Trud, 5 December 1971: 1; Pravda, 14 December 1971: 3; Soviet Weekly, 18 March 1972: 4; Geologiya i perspektivy neftega-zonosnosti Sovetskoi Arktiki (The geology of potential oil and gas bearing areas in the Soviet Arctic), Leningrad 1972.

19. Energeticheskie resursy, op. cit.: 291; Neftyanoe khoz-yaistvo no. 12, December 1972: 3-4.

20. OECD Observer no. 54, October 1971: 30-31.

21. Pravda, 19 October 1961: 3.

22. Neftyanik no. 7, July 1971: 1; Neftyanik no. 8, August 1971: 4.

23. Energeticheskie resursy, op. cit.: 203-208; Driatskaya, Z. V., et al., Nefti SSSR (Oils of the USSR), Moscow 1971: Vol 1, 437.

24. Mingareev, R. Sh., Valikhanov, A. V., "Tatarskaya ASSR," Neftyanoe khozyaistvo no. 3, March 1971: 10-13; Mirolyubova, E., "Pervyi milliard tonn Tatarskoi nefti" (First 1000 million tons of Tatar oil), Neftyanik no. 8, August 1971: 4-5.

25. Karzanov, Yu., "Goroda gde my zhivem" (The towns where we live), Neftyanik no. 3, March 1971: 32.

26. Energeticheskie resursy, op. cit.: 208-12; Driatskaya, ° Vol. 1 op. cit.: 292-295.

27. Neftyanik no. 3, March 1971: 8; Stolyarov, E. V., "Bash-kiriya, krai nefti" (Bashkiria, land of oil), Pravda, 28 June 1971: 2; Latypov, M., Neftyanik no. 1, January 1972: 6-7; Sotsialisticheskaya industriya, 29 April 1973: 2.

28. Energeticheskie resursy, op. cit.: 213-18; Orlov, V. P. Grigorashchenki, G.I., "Kuibyshevskaya oblast," Neftyanoe khozyaistvo no. 3, March 1971: 21-24; Driatskaya, vol. 2, op. cit.: 13-15.

29. Ibid.: 270-73; Energeticheskie resursy, op. cit.: 219-20, 358.

30. Ibid.: 218, 359.

31. Ibid.: 201-203; Alekseev, P. D., "Orenburgskaya oblast," Neftyanoe khozyaistvo no. 3, March 1971: 49-53; Driatskaya, Vol. 2, op. cit.: 151-53.

32. Energeticheskie resursy, op. cit.: 195-200; Maltsev, N. A., "Permskaya oblast," Neftyanoe khozyaistvo no. 3, March 1971: 33-36.

33. Energeticheskie resursy, op. cit.: 200; Driatskaya, Vol. 1, op. cit.: 251-53.

34. Energeticheskie resursy, op. cit.: 227-30; Nazarov, V. B., "Checheno-Ingushskaya ASSR," Neftyanoe khozyaistov no. 3, March 1971: 24-27;

35. Energeticheskie resursy, op. cit.: 224-27; Khomyakov, A. A., Bragin, V. A., "Krasnodarskii krai," Neftyanoe khozyaistvo no. 3, March 1971: 56-59.

36. Energeticheskie resursy, op. cit.: 221-24; Shelomentsev, G. I., "Stavropolskii krai," Neftyanoe khozyaistvo no. 3, March 1971, 54-56.

37. Energeticheskie resursy, op. cit.: 230-31.

38. Ibid.: 190-94; Driatskaya, op. cit.: 19-21; Shashin, V. D., Neftyanoe khozyaistvo no. 3, March 1971: 8.

39. Muravlenki, V. I., "Zapadnaya Sibir," Neftyanoe khozyaistvo no. 3, March 1971: 15-20; Nesterov, I. I., et al., Neftyanye i gazovye mestorozhdeniya Zapadnoi Sibiri, Moscow 1971; Energeticheskie resursy, op. cit., 232-36; Shashin, V. D., "Vnimanie—Zapadnaya Sibir" (Attention—Western Siberia), Neftyanik no. 4, April 1970: 21-22; Lavrov, N., "Tyumenskie besedy" (Tyumen conversations), Neftyanik no. 1, January 1971: 2-7; Efremov, E. P., et al., "Razvitie neftyanoi promyshlennosti Zapadnoi Sibiri v 1964-1970 gg" (Development of the oil industry in Western Siberia 1964-1970), Neftyanik no. 7, July 1971: 7-10; Shibanov, V., "Burovaya ustanovka na vozdushnoi podushke" (Drilling rigs on hovercrafts), Neftyanik no. 2, February 1972: 20-21; Neftyanik no. 12, December 1972: 6-8; Pravda, 23 May 1971: 2, 4 March 1972: 1-2, 2 July 1972: 2, 13 July 1972: 2, 17 July 1972: 1, 18 July 1972: 1, 24 July 1972: 2, and 9 August 1972: 1 ; Sotsialisticheskaya industriya, 27 January 1971: 2, 19 January 1973: 1, and 30 March 1973: 1; Izvestiya, 20 March 1973: 1; Komsomolskaya pravda, 1 April 1973: 2; Trud, 11 April, 1973: 1.

40. Energeticheskie resursy, op. cit.: 236-38.

41. Ibid.: 238-39.

42. Ibid.: 249-56; Pravda, 29 March 1971: 2.

43. Amirov, A. D., et al, "Azerbaidzhanskaya SSR," Neftyanoe khozyaistvo no. 3, March 1971: 28-32; Gadzhiev, B., "Morskaya neft dolzhna byt deshevle" (Sea oil should be cheaper), Neftyanik no. 5,

May 1971: 4-6, 24-27; Sotsialisticheskaya industriya, 19 January 1971: 1.

44. Energeticheskie resursy, op. cit.: 263-66; Dadashev, Sh. A., et al., "Turkmenskaya SSR," Neftyanoe khozyaistvo no. 3, March 1971: 36-42.

45. Energeticheskie resursy, op. cit.: 257-63; Utebaev, S. U., "Kazakhskaya SSR," Neftyanoe khozyaistvo no. 3, March 1971: 45-49; Nikolaeva, M. N., "Na poluostrove Mangyshlak," Stroitelstvo trubo-provodov no. 12, December 1972: 26-27.

46. Energeticheskie resursy, op. cit.: 269-70; Ismailov, A. I., "Uzbekskaya SSR," Neftyanoe khozyaistvo no. 3, March, 1971: 60-62.

47. Energeticheskie resursy, op. cit.: 266-71; Narodnoe khozyaistvo SSSR v 1969 godu, Moscow 1970: 197; SSSR i soyuznye respubliki v 1972 godu.

48. Energeticheskie resursy, op. cit.: 240-46; Potyukaev, M. A., "Ukrainskaya SSR," Neftyanoe khozyaistvo no. 3, March 1971: 42-45.

49. Energeticheskie resursy, op. cit., 247-49.

50. SSSR i soyuznye respubliki v 1970 godu, Moscow 1971; BBC Summary of World Broadcasts: USSR, Weekly Economic Report: SU/W673/A/7, SU/W 682/A/8; Izvestiya, 6 April 1973: 3.

51. Energeticheskie resursy, op. cit.: 330-42; Melnikov, op. cit.: 47-60; Pravda, 15 July 1972: 2, 11 August 1972: 1; Sotsia-listicheskaya industriya, 26 February 1971: 1.

52. Shashin, N. V., "Direktivy XXIII sezda Partii vypolneny" (Directives of the 23rd Congress are fulfilled), Neftyanoe khozyaistvo no. 3, March 1971: 3-4; Neftyanik no. 1, January 1971: 5; Neftyanik no. 3, March 1971: 10-11; Neftyanik no. 6, June 1971: 35; Vneshnyaya torgovlya SSSR 3a 1971, op. cit.: 96; 3a 1972 god: 95.

53. Melnikov, op. cit.: 61-62.

54. Ibid.: 64.

55. Narodnoe khozyaistvo SSSR v 1969 godu, Moscow 1970: 445-446.

56. Melnikov, op. cit.: 65; Transport i khranenie neftii nefte-produktov (Transport and storage of oil and products) no. 11, November 1969: 6-8.

57. Yufin, V. A., et al., Truboprovodnyi transport, Moscow 1970: 18-25; Kostrin, K. V., "70 let otechestvennomu truboprovodos-troeniyu," Stroitelstvo truboprovodov no. 1, January 1970: 8-9; Boltalina, E. F., "Druzhba," Stroitelstvo truboprovodov no. 4, April 1970: 32-33; Karapetyan, A. G., "Novaya programma rabot v Zapadnoi Sibiri" (New program of operations in Western Siberia), Stroitelstvo truboprovodov no. 4, April 1970: 23-24; Vasiliev, N. P., et al., "Sooruzhenie nefteprovoda Aleksandrovskoe Anzhero-Sudzhensk," Stroitelstvo truboprovodov no. 1 January 1971: 13-14; Ekonomicheskay

gazeta no. 18, April 1972: 24; Sotsialisticheskaya industriya, 20 February 1971: 1; Pravda, 2 January 1972: 1, 31 January 1972: 1, 4 March 1972: 1-2, 15 February 1973: 1, 2 April 1973: 1; Trud, 29 March 1973: 1, 1 May 1973: 3.

4

Coal has at least one major long-term advantage over oil and gas: there are considerably more reserves of coal in the world than of any other fossil fuel. While oil and gas account for about 3 percent each and oil shales for 6 percent, coal accounts for as much as 88 percent.[1]

In recent years coal has had less prominence in the Soviet press, which has concentrated on the more rapidly developing oil and gas industries. Yet it was only in 1968 that coal was replaced by oil as the primary source of energy, and coal will still be providing about 30 percent of the Soviet fuel requirement by 1975.

The USSR has continued to increase its annual output of coal, and now claims to be the world's major producing country. Coal is not only important in the Soviet energy balance; in 1970 about 180 different chemical products owed their origin to this valuable raw material.

HISTORICAL BACKGROUND

The Russian coal industry began its development in the reign of Peter the Great, with the formation in 1700 of a Department of Mining, which became one of the nine governmental colleges (the "Berg-kollegiya") in 1719. Three major coal basins were discovered soon after, the Donets basin (Donbass) in 1721 and the Moscow and Kuznetsk basins in 1722. A certain amount of capital was invested

Some recent data have been added to this chapter from the BBC monitoring service, USSR Weekly Economic Report (coal), 1972-73.

by merchants in the new coal mines, but most of the mines depended heavily on state subsidies and protective tariffs. Most of the miners were serfs or vagrants conscripted by the state.

In the 18th century output remained insignificant, although it began to increase with the construction in 1795 of the Lugansk iron works, which was fueled by coking coal from the Donbass. Experiments with Kuzbass coal were later tried in the Altaisk iron works.

The first large blast furnace began production in the Donbass in 1871, and by 1900 there were twelve metals plants in operation, requiring a sharp increase in the output of hard caking coals.

The Russian railway network grew from 3,800 km. in 1865 to 20,000 km. in 1890 and 58,500 km. in 1913. This not only made it possible for Donbass coal to be transported to a much wider market, but also created a steadily rising demand for locomotive fuel.

The early development of the Russian coal industry is shown in Table 4.1.

In spite of significant increases in the decades before the First World War, Russian coal production still did not reflect the vast reserves available, and in 1913 it actually formed only 2.5 percent of the total world output. The irrational geographical distribution of the coal industry made it economically more expedient for St. Petersburg consumers to import coal from abroad; in 1913 as much as 8.7 million tons of coal were shipped into Russia. At this time the Donbass was

TABLE 4.1

Coal Production in Russia, 1863-1913
(in thousands of tons)

Basin or Region	1863	1873	1883	1893	1903	1913
RUSSIA (total)	159	833	2,287	4,447	13,079	29,117
Donets basin	105	616	1756	3929	11583	25288
Moscow basin	22	151	372	179	218	300
Ural basins	12	16	126	261	491	1217
Kuznetsk and Minusinsk basins	10	13	27	18	249	799
East Siberia and the Far East	10	2	6	12	470	1195

Source: "Ugolnaya promyshlennost," Bolshaya Sovetskaya Entsiklopediya, Moscow 1949-57, Vol. 43: 627.

producing 86.9 percent of the total national output and 95.5 percent of the output of European Russia.

Extraction methods were primitive. In 1913 the Donbass had so few cutting machines that only 1.7 percent of the coal output was mechanized. The average output per miner for the year was only 153 tons, compared with 287 tons in Germany and 264 tons in Britain. Wages were low, and the number of accidents was high.

In 1917 some 59 percent of the Russian coal industry was owned by foreign investors. Of the 25 large companies with stock in Russian hard coal production, 19 were either French or Belgian. Although output rose to 34.5 million tons in 1916, the requirements of the war economy could not be met. Thirty blast furnaces had to be closed down, and iron and steel production in the south dropped by half.[2]

During the Civil War, which followed the 1917 revolution, Russia suffered from a crippling fuel shortage. The Donbass was occupied by the Germans in the spring of 1918 and then by General Denikin's White forces until the end of 1919. Many of the miners fought in the ranks of the Red Army under Voroshilov on the Tsaritsyn front, and there was much bitter fighting in the Donbass region itself. The Kuznetsk basin and the coal fields of the Urals, Siberia, and the Far East also suffered destruction and the loss of skilled labor before the war finally ended in 1921. Output had dropped to a mere 8.7 million tons in 1920.

The new Soviet government took several immediate measures to reconstruct the coal industry. As early as 7 December 1917 a central fuel committee (Glavtop) had been established; in August 1919 a decree was passed on the reconstruction of coal mines in the areas being brought under Soviet control, and in April 1920 the general mobilization of all workmen in the mining industry was announced. The GOELRO plan of December 1920 called attention to the necessity of regaining prewar production levels in the mines of the Donbass and Urals, in order to supply fuel for the new power stations.

Output began to increase, rising from 11.3 million tons in 1922 to 32.3 million in 1927. Many new coal dressing factories were built to provide high-quality coking fuel for the expanding metals industry. In 1932 steps were taken to raise the standard of technology in the Donbass by importing the latest coal extraction machinery from abroad and by beginning the production of coal cutting and transporting equipment in Gorlovka, Kharkov, and Donetsk factories.

The Donbass was regarded almost as the cradle of Soviet industrialization, the foundation on which the metals, energy, and chemicals industries were to be built. In 1928 it was still providing 78 percent of the nation's coal, in spite of considerable expansion east of the Urals.

In October 1932 the discovery of the first coal deposit of the Pechora basin, near Vorkuta, was announced.

Other important developments were the construction of the Kashira and other power stations, based on local brown coal from the Moscow basin, and the formation of the Ural-Kuznetsk iron and coal complex, known as the Magnitogorsk-Kuznetsk Combine. Steel mills were built both at the iron ore deposits in the southern Urals and near the Kuznetsk coal mines, thus fully utilizing railway cars by transporting iron ore eastward and coking coal westward.

In the period from 1933 to 1937, the second Five Year Plan, 25 large, well-equipped mines were built in the Kuzbass. The exploitation of the Karaganda basin commenced at about the same time. In 1931 work began on the first twenty mines, with another five being constructed the following year. By 1940 output at this third Soviet coal base had reached 6 million tons.

The use of machinery in the mines increased rapidly. In 1928 the Soviet coal industry had only 549 cutting machines; their number rose to 1,473 in 1932 and to 3,421 in 1940.

It was in the 1930s that the famous Stakhanovite movement developed. In August 1935, with the help of two men to install pit-props, Aleksei Stakhanov succeeded in cutting 102 tons of coal in 5 hours 45 minutes, over 14 times the shift norm of 7 tons. In September he increased his record to 227 tons in one shift, only to be surpassed by Nikolai Izotov (240 tons) and Fedor Artyukhov (311 tons). These feats were widely publicized, and the press urged workers in all other branches of industry to emulate these first "innovators." How Stakhanov's less energetic colleagues felt about the new movement is not recorded, but the average productivity per man per shift certainly increased greatly over the following years.*

Production of coal in the decade before the Second World War increased very rapidly indeed. In 1928 output had only matched the 1916 figure of 35.5 million tons; by 1932 it had risen to 64.4 million, and only by 1940 had it jumped to 165.9 million.

Open pit mining was being developed. In 1940 almost 4 percent of the total coal production was by this more economical method, which was practised mainly in the Urals (Korkino, Bogoslovsk), East Siberia (Cheremkhovo), and the Far East (Raichikhinsk). Many underground mines were also being constructed in new fields throughout the USSR; in the period from 1929 to 1940 over 400 new mines were brought

---

*Thirty-five years later, on "Miners' Day" (30 August 1970), Stakhanov was still front-page news, when his speech encouraging a new generation of Soviet miners was fully reported, and in 1973 he was reported to be writing his memoirs. (Izvestiya, 6 March 1973.)

into operation. The first coal combines were appearing, and approximately 90 percent of coal mining had been mechanized.

In spite of considerable expansion in the eastern coal fields, the Soviet energy balance was still much too vulnerable to invasion from the West. By the end of 1941 both the Donbass and the Moscow basin had been occupied by the Germans, and coal production in the USSR dropped to less than half the prewar level. The effect of the war on the Soviet coal industry is indicated in Table 4.2.

Some 350 mines were destroyed in the Donbass, and almost all the surface installations were left in ruins; by 1945, despite all-out efforts to reconstruct the basin, production was only 38.4 million tons, compared to 85.5 million tons in 1940. Damage to the Moscow basin was less severe, and by 1945 output was more than double the prewar figure. The construction of new mines in the Kuznetsk and Karaganda basins was intensified during the war, often using equipment removed from the occupied territories. Production in the Urals fields expanded rapidly, from 11.7 million tons in 1940 to 25.1 million in 1945. The completion of the railway from Vorkuta to the Center allowed a flow of coking coal from the Pechora basin to reach the metals plants in the south. The amount of coal mined by opencast methods tripled during the war years.

Many miners left for the front, having been enlisted in the Miners Regiments, and were frequently replaced by women and youths. The organization and sheer human effort involved in maintaining fuel

TABLE 4.2

Soviet Coal Production, by Basin, 1940-45
(in millions of tons)

Basin	1940	1941	1942	1943	1944	1945
USSR	165.9	151.4	75.5	93.1	121.5	149.3
Donets	85.5	66.0	3.8	4.1	20.3	38.4
Moscow	9.9	9.5	8.6	14.4	17.6	20.2
Pechora	0.26	0.3	0.75	1.7	2.5	3.3
Kuznetsk	21.1	25.1	21.0	24.9	27.1	30.0
Karaganda	6.3	7.2	6.9	9.6	10.8	11.3
Urals fields	11.7	14.1	15.8	20.5	22.9	25.1

Source: Bratchenko, B. F., et al., Ugolnaya promyshlennost SSSR 1917-1967, Moscow 1969: 14.

supplies to industry during the war is in many ways as admirable as the final military victory over Nazi Germany.

The Donbass was liberated early in 1943, and reconstruction began immediately; often makeshift methods had to be used, as, for example, when railway locomotives were adapted to generate electric power. Although by 1949 the prewar level of output was attained, the Donbass has never regained its former excessive share of total Soviet production, since the expansion to other areas has been maintained. The Donbass, however, has still preserved the honor of being the major single producer of coal in the USSR.

Changes in the geographical distribution of Soviet coal production since 1917 are shown in Table 4.3 and Figure 4.1.

By the end of the war the share of the European part of the USSR had dropped to 42 percent, while the Urals fields were contributing 17 percent and the new fields in the eastern areas as much as 41 percent. By 1972 the latter were accounting for 46 percent of coal production.

Energy for the rapid expansion of Soviet industry in the postwar years was supplied mainly by coal, its share in the fuel balance rising from 62 percent in 1945 to 66 percent in 1950, when 261.1 million tons were produced. Although the increased utilization of the more economic fuels, oil and gas, has meant that the role of coal has become less dominant than before, actual output has continued to rise.

From 1945 to 1960 coal output increased an average of 24 million tons a year; the rate of increase in the 1960s fluctuated to a much greater extent but still averaged over 10 million tons a year and is expected to be over 15 million tons a year in the 1970s; over 695 million tons are planned to be produced in 1975. (See Table 4.4 and Figure 4.2.)

In 1972 some 655 million tons of coal were produced in the USSR; of this, 170 tons were suitable for coking. Most of this vast output was from underground mines, but 182 million tons (29 percent) were extracted by opencast methods.

Research on Soviet economic-history is often rendered particularly baffling by a confusing diversity of organization titles, which makes it difficult to know which part of which industry is being referred to in certain statements. An excellent example of this is provided by the coal industry, which has frequently changed its name since the formation of Soyuzugol (Union Coal) in November 1929. In January 1939 it was decided to unite the administration of the various fuel industries under a single People's Commissariat for the Fuel Industry of the USSR (Narodnyi komissariat toplivnoi promyshlennosti SSSR), which was then in October divided into two separate Commissariats for the oil and coal industries. Then, in January 1946, the Commissariat for the coal industry was subdivided into two People's Commissariats (Narkom), with one for the western regions and one for the

127

TABLE 4.3

Changes in the Geographical Distribution of the Soviet Coal Industry, 1917-70
(in percentage)

	1917	1928	1940	1945	1955	1960	1965	1967	1970
Production by region (as percentage of total)									
European USSR	81.6	80.5	63.9	42.3	53.0	52.5	50.0	48.7	49.4
Donets Basin	79.0	77.0	56.8	25.6	36.1	36.9	35.7	34.8	34.6
Pechora Basin	–	–	0.1	2.2	3.6	3.5	3.1	3.1	3.4
Moscow Basin	2.2	3.2	6.1	13.6	10.0	8.4	7.1	6.6	5.8
Lvov-Volynsk Basin	–	–	–	–	–	0.7	1.7	1.7	2.0
Others	0.4	0.3	0.9	0.9	3.3	3.0	2.4	2.5	3.4
Urals	4.9	5.6	7.2	17.2	12.0	11.5	10.7	9.6	7.4
Eastern USSR	13.5	13.9	28.9	40.5	35.0	36.0	39.3	41.7	43.2
Kuznetsk Basin	4.2	7.4	13.6	20.1	15.0	16.5	16.8	17.5	17.7
Karaganda Basin	–	–	3.8	7.6	6.3	5.1	5.3	5.2	6.2
Ekibastuz Basin	–	–	–	–	0.6	1.2	2.5	2.9	3.6
Eastern Siberia	6.4	2.8	5.6	6.2	6.6	7.1	8.1	9.4	9.1
Far East	1.9	3.0	4.3	5.3	4.5	4.5	4.9	4.9	4.7
Central Asia	1.0	0.7	1.0	0.9	1.5	1.5	1.6	1.6	1.3
Others	–	–	0.6	0.4	0.3	1.1	0.1	0.2	0.6
Total production in millions of tons	35	35.5	165.9	149.3	389.9	509.6	577.7	595.2	624.1

Sources: Bratchenko, B. F., et al., Ugolnaya promyshlennost SSSR 1917-1967, Moscow 1969: 15; Narodnoe khozyaistvo SSSR, 1922-1972, Moscow 1972: 142; Table 4.15.

FIGURE 4.1

Changes in the Geographical Distribution of Coal Industry, 1917-1970

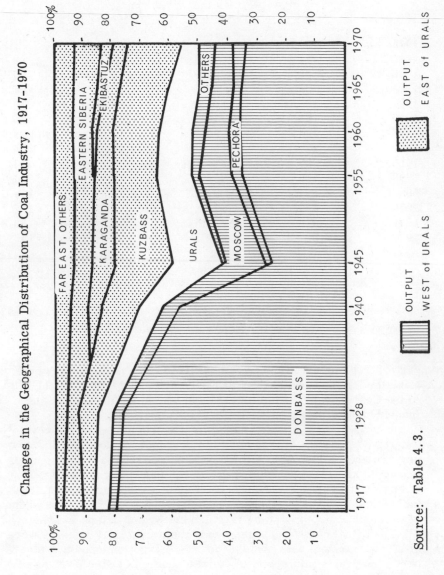

Source: Table 4.3.

TABLE 4.4

Production of Coal in USSR, 1913-72
(in millions of tons)

Year	Total Production	Hard Coal		Brown Coal
		Total	Anthracite	
1913	29.2	28.0	4.8	1.2
1922	11.3	9.3	2.2	2.0
1928	35.5	32.5	8.0	3.1
1932	64.4	57.5	18.1	6.9
1937	128.0	109.9	28.0	18.1
1940	165.9	140.0	35.7	25.9
1945	149.3	99.4	16.9	49.9
1950	261.1	185.2	40.2	75.9
1955	389.9	276.6	57.8	113.3
1960	509.6	374.9	74.1	134.7
1965	577.7	427.9	76.5	149.9
1970	624.1	476.4	75.8	147.7
1971	640.9	487.5	75.8	153.3
1972	655.2	499.5*	75.4*	155.7*

*Figures for USSR and Ukrainian Coal Ministries.

Sources: Narodnoe khozyaistvo SSSR, 1922-1972, Moscow 1972: 164; Ugol no. 4, April 1973: 72.

eastern regions. Next, in December 1948, both of the above, plus the Ministry for the Construction of Fuel Enterprises, became the Ministry of the Coal Industry of the USSR, from which the Ministry of the Coal Industry of the Ukrainian SSR was separated in April 1954.

Apart from a simple change in terminology (after the revolution the word "ministry" had seemed too bourgeois to describe the new governmental divisions), new titles often reflected a more complex economic structure, where loyalty to the Party was not in itself enough to ensure the successful administration of an industry. Specialists had to be appointed as heads of the important branches of the economy, yet overall control had to be retained by the politicians in Moscow. When political control changed hands the ministerial domains were often reorganized, with some ministers being dismissed and others becoming deputy ministers. After the death of Stalin there were several new reorganizations of ministries by those in power. In 1957

FIGURE 4.2

Expansion of Coal Production in USSR, 1913-1973

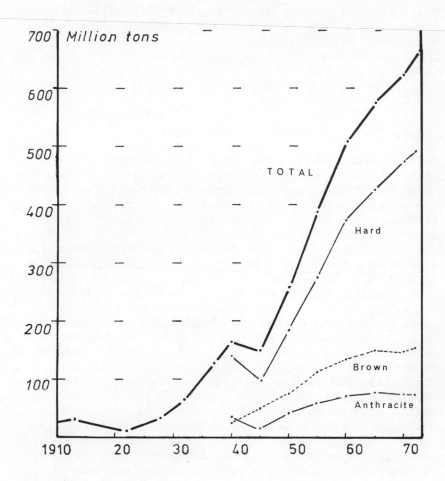

Source: Table 4.4.

the various ministries were abolished in favor of a regional structure in order to correct the previous lack of cooperation between ministries, as each attempted to be self-sufficient. The state committees that followed were subject to new abuses, and with the advent of Brezhnev and Kosygin in 1964 the ministries were restored; the Ministry of the Coal Industry reappeared in November 1965. In the 1970s its administration is again subdivided, with a separate Ministry of the Coal Industry of the Ukrainian SSR.[3]

PRESENT AND FUTURE DEVELOPMENT

Exploration and Location of Reserves

Although prospecting for coal involves the same basic methods as for oil and gas, somewhat different criteria are taken into consideration when grouping reserves according to their degree of exploration. The thickness, extent, depth, and quality of the coal seams must be carefully considered when deciding whether exploitation will be economically viable.

Deposits that fulfill industrial requirements in quality and are suitable for extraction at the present level of technology are known as "balance resources" (balansovye resursy). Deposits that are found in thin seams or are of poor quality, or that require complex processing, are therefore not at present suitable for exploitation and are known as "extra-balance resources" (zabalansovye resursy).

As with oil and gas, several coal deposits (zalezh) may be found in one field (mestorozhdenie). There is no precise distinction drawn between a coal field and a coal basin (bassein), although the latter term is generally reserved for the largest coal-bearing areas, sometimes covering tens or hundreds of thousands of square kilometers and containing several fields.

Coal reserves are also classed in the categories A (fully explored), B (main features known), C_1 (general characteristics known), and C_2 (by preliminary estimation). The sum of these categories plus the predicted reserves (prognoznye zapasy) are known as the total geological reserves (obshchegeologicheskie zapasy) or as the potential reserves (potentsialnye zapasy).

Mines are not normally constructed until deposits are known to contain explored balance resources of categories A, B, or C_1. These categories are normally what is meant by industrial or commercial reserves (promyshlennye zapasy); that is, those explored well enough to be ready for exploitation.

Exploration for coal resources in the USSR is carried out by geologists in the Ministry of Geology and Conservation of Natural Resources (MGION), the Ministry of the Coal Industry of the USSR, or the Ministry of the Coal Industry of the Ukrainian SSR.

The explored balance coal reserves of the USSR, according to calculations from the VFG (All-Union Geological Fund) in 1966, amounted to 237,200 million tons $(A + B + C_1)$ and 235,000 million tons (C_2), a total of 472,200 million tons. In 1970, after further exploration, it was estimated that a more accurate assessment of $A + B + C_1 + C_2$ reserves would be about 2,200,000 million tons. Proven reserves are still being expanded by constant exploration.[4]

The VFG of the Ministry of Geology and Conservation of Natural Resources supplies a figure of 8,669,000 million tons for the total geological coal resources of the USSR, based on the norms established by the 1913 International Geological Congress. These stipulated that calculations should be based on hard coal seams over .3 m. thick and on brown coal (lignite) seams over .6 m. thick, lying at depths of less than 1,800 m. If we assume a figure of 16,000,000 million tons for the total geological coal reserves of the world, the USSR can claim as much as 55 percent of the earth's coal.

According to the norms now accepted in the USSR, total geological balance reserves amount to 7,765,300 million tons, of which 4,504,500 million is hard (kamennyi) coal and 3,260,800 million is brown (buryi). This compares well with figures for the United States; according to U.S. Geological Survey estimates, as of January 1967 there were some 3,200,000 million tons of coal at levels of 0 to 6,000 feet.

In 1954 the total geological reserves of the United States were calculated at 1,723,000 million tons. Working from the same norms — hard coal in seams over .6 m. thick at depths of less than 1,200 m. overburden and brown coal in seams over .9 m. thick at depths of less than 500 m. overburden—Soviet experts estimated that the USSR had some 5,521,000 million tons, more than three times the reserves of the United States.[5]

Statistics from the 1968 World Power Conference Survey of Energy Resources include coal reserves at a maximum depth of over-burden of 1,200 m. and a minimum seam thickness of 30 cm. The USSR accounted for 32 percent of the total measured hard coal reserves and 39 percent of the measured brown coal reserves. The equivalent figures for North America were 25 percent and 8 percent. The proportion of the world's total reserves allotted to the USSR was even higher: 63 percent of a total of 8,816,600 million tons, compared to the 18 percent share of North America.

Since estimates of the world's mineral reserves change every year with the results of new exploration, and since calculations vary according to the norms accepted, such comparisons are no more than approximate. It is certain, however, that the USSR has sufficient coal reserves to maintain increased production for centuries to come and that its ratio of reserves to output is healthier than the ratios for most large industrial states.

A particularly detailed and comprehensive survey of Soviet coal reserves was initiated in 1954 and published in 1956. The total geological balance reserves, 7,765,300 million tons, were divided according to depths as follows: less than 300 m., 2,088,800 million tons (26.9 percent); from 301 to 600 m., 1,599,600 million tons (20.6 percent); from 601 to 1,200 m., 2,540,700 million tons (32.7 percent); from 1,201 to 1,800 m., 1,536,200 million tons (19.8 percent). About half of these reserves are therefore at depths that are easily mined, and indeed almost 70,000 million tons could be mined by the most economical opencast methods.

Some 93 percent of the total geological balance reserves; 82 percent of the reserves explored to categories $A + B + C_1 + C_2$; and 76 percent of the reserves explored to categories $A + B + C_1$ lie in the area east of the Urals.[6]

The regional distribution of balance reserves is shown in Table 4.5. Soviet coal reserves are therefore concentrated in the RSFSR, the Ukraine, and Kazakhstan, as may be seen more clearly from Table 4.6.

Table 4.7 breaks down the statistics for the reserves of coal in the USSR according to their distribution among the main basins and fields, indicating the quantities of both hard (kamennyi) and brown (buryi) coal in each. The difference between the potential reserves and the proved reserves is particularly noticeable in the case of the poorly explored coal basins in the east of the USSR, such as the Kansko-Achinsk, Tunguska, Lena, and Taimyr basins. The balance reserves for the main coal basins are given as a percentage of the USSR total reserves in Table 4.8. The location of the main basins is shown on Map 4.1. Since this detailed and comprehensive survey was compiled in 1966, new data has become available for some individual fields and basins, and this is included in the descriptions that follow below. In general, however, coal reserve figures have changed very little in recent years, compared with those for natural gas.

The Soviet coal classification system is different from that used elsewhere, and a list of commonly used terminology for the various grades of coal is included in the glossary. The total balance reserves have been divided by grades as follows: anthracite and lean coal (A + T), 11.6 percent; caking coals (Zh + K + OS), 21.8 percent; gas coal (G), 8.6 percent; longflame cannel coal (D), 16 percent; longflame brown coal (DB), 5.1 percent; brown coal (B), 36.9 percent.[7]

TABLE 4.5

Regional Distribution of Balance Coal Reserves
(in thousands of millions of tons)

Region	Total Balance Reserves, 1956	Explored $(A+B+C_1)$ 1966
RSFSR	7,452.1	170.0
Northwest	263.8	6.9
Center	17.5	4.4
North Caucasus	43.7	5.3
Ural	35.1	3.0
Western Siberia	849.9	83.1
Eastern Siberia	3,589.7	55.8
Far East	2,652.4	11.5
Ukrainian SSR	151.7	37.2
Donets-Dnepr	148.5	36.4
Southwest	3.2	0.8
Transcaucasia	0.7	0.4
Central Asia	38.3	4.3
Kazakh SSR	122.5	25.3
Total for USSR	7,765.3	237.2

Source: Energeticheskie resursy SSSR: Toplivno-energeticheskie resursy, Moscow 1968: 65.

Since the 1950s coal exploration has taken second place to oil and gas when well-qualified geologists are allocated. At a conference held in Kemerovo in October 1971, complaints were voiced about the poor quality of the equipment provided to geologists and about the shortage of trained personnel. Out of the 1,097 men who were employed by the coal industry of the USSR in some form of geological work, only 30 percent had had a higher education specializing in geology; insufficient attention was being paid to raising the qualifications of mine geologists.

For the 1971-75 period, however, the position has been somewhat improved. Some 95 percent more resources have been assigned to coal exploration than in the previous five years. By 1975 the exploration of the 157 fields now being exploited is to be completed, and further work is to be carried out on the 90 mines being reconstructed and on the 9 mines being built. This entails the drilling of

TABLE 4.6

Distribution of Coal Reserves, by Republic
(in percentage)

Republic	Geological (Balance) Reserves	Industrial Reserves	
		$A + B + C_1 + C_2$	$A + B + C_1$
RSFSR	96.0	80.3	71.7
Ukraine	1.9	9.9	15.7
Kazakhstan	1.6	8.3	10.6
Central Asian republics	0.49	1.4	1.8
Caucasian republics	0.01	0.1	0.2

Source: Energeticheskie resursy SSSR: Toplivno-energeticheskie resursy, Moscow 1968: 66.

7.4 million m. of test bores and the sinking of 231,000 linear m. of exploratory mines.[8]

Major Basins and Fields

Although a relatively small share of the Soviet coal reserves are contained in European USSR, the requirements of industry and of huge conurbations have given the coal deposits west of the Urals particular importance.

The Donets Basin (Donbass). Covers an area of over 60,000 sq. km., stretching over 640 km. east and west of the Don through the Donetsk, Lugansk, Dnepropetrovsk, and Rostov oblasts. Reserves of categories $A + B + C_1 + C_2$ were calculated in 1970 at about 90,000 million tons. Some 27 percent is anthracite and 17 percent coking coal (1964). Total balance reserves are given by grades in Table 4.9. There are over 140 seams, mainly shallow, those being worked averaging .93 m. There are some 550 mines in operation at an average depth of 340 m.; only 8 percent of the mines reach depths from 600 to 900 m. About 42 percent of the balance reserves in those mines is coking coals; most of the remainder is high-quality energy coal, of which over 30 percent is anthracite. The ash content of mined coal is low, but most of the coal is of a medium or high sulphur level.

TABLE 4.7

Hard and Brown Coal Reserves, by Basin
(in thousands of millions of tons)

Basin	Total Geological Reserves (no deeper than 1,800 m.)		Balance Reserves Proved to $A + B + C_1 + C_2$ (1965)	
	Hard	Brown	Hard	Brown
Donets	219.1	21.5	53.0	—
Pechora	344.5	—	13.8	0.5
Moscow	—	24.3	—	6.1
Lvov-Volynsk	1.7	—	0.7	—
Dnepr	—	4.2	—	2.5
Kizel	1.1	—	0.6	—
Chelyabinsk	—	1.6	—	0.9
South Ural	—	1.8	—	1.3
Karaganda	50.0	1.2	13.6	0.4
Ekibastuz	12.2	—	9.4	—
Maikyuben	—	21.0	—	4.9
Turgai	—	36.5	—	6.7
Kuznetsk	849.4	55.9	167.2	23.3
Kansko-Achinsk	1.8	1,218.5	1.6	83.4
Minusinsk	36.9	—	2.4	—
Irkutsk	84.7	4.2	19.1	2.8
Tunguska	1,553.2	190.8	3.0	0.9
Lena	1,141.7	1,505.5	1.7	1.7
South Yakut	40.0	—	5.2	—
Transbaikal	1.2	7.2	1.3	5.7
Taimyr	555.4	28.1	0.7	—
Bureya	25.0	—	2.0	—
Suchan	1.4	—	0.4	—
Central Asia	26.6	14.2	2.6	4.3
USSR	5,182.5	3,487.0	311.2	157.4

Source: Adapted from Melnikov, N. V., Mineralnoe toplivo, Moscow 1971: 22.

TABLE 4.8

Distribution of Coal Reserves
(in percentage)

Area or Basin	Total Balance Reserves (1956)	Explored Reserves (1966)	
		$A + B + C_1$	$A + B + C_1 + C_2$
European USSR	6.53	22.8	16.8
Donets	2.44	16.5	11.6
Moscow	0.23	1.9	1.3
Pechora	3.37	2.9	3.0
Lvov-Volynsk	0.02	0.3	0.2
Dnepr	0.05	1.1	0.6
Georgia	0.01	0.1	0.1
Others	0.41	—	—
Urals	0.09	1.3	0.8
Kizel	0.013	0.2	0.2
Chelyabinsk	0.020	0.4	0.2
South Urals	0.020	0.6	0.3
Sverdlov oblast	0.013	0.1	0.1
Others	0.024	—	—
East of Urals	93.38	75.9	82.4
Kuznetsk	10.35	22.1	41.0
Karaganda	0.60	3.2	2.9
Turgai	0.46	2.6	1.4
Maikyuben	0.17	0.7	1.1
Ekibastuz	0.14	3.1	1.5
Other parts of Kazakhstan	0.21	1.0	1.3
Central Asia	0.49	1.8	1.5
Minusinsk	0.47	1.1	0.6
Kansko-Achinsk	15.52	28.7	18.6
Ulukhem	0.13	0.3	0.3
Irkutsk	0.86	2.9	4.6
Transbaikal	0.10	2.3	1.6
Tunguska	19.52	0.8	0.8
Taimyr	6.59	0.1	0.2
Lena	31.14	1.0	0.7
South Yakut	0.51	0.9	1.1
Bureya	0.37	0.3	0.4
Uglov	0.01	0.1	0.1
Suchan	0.02	0.1	0.1
Verkhne-Suifun	0.02	—	0.1
Magadan oblast	0.05	0.5	0.9
Sakhalin	0.25	0.9	0.8
Others	5.42	1.4	1.4
USSR (Actual reserves in thousands of millions of tons)	7.765.3	237.2	472.2

Source: Adapted from Energeticheskie resursy SSSR: Toplivno-energeti-cheskie resursy, Moscow 1968: 64.

TABLE 4.9

Donets Balance Reserves, by Grade

Grade	Total Reserves (in thousands of millions of tons)	(in percentage)
B + DB	20.9	11.0
D	24.2	12.7
G	54.3	28.5
Zh	10.9	5.8
K + KZh	7.7	4.1
OS	6.6	3.5
T	14.7	7.8
PA + A	50.7	26.6
Total	190.0	100

Source: Energeticheskie resursy SSSR: Toplivno-energeti-cheskie resursy, Moscow 1968: 71.

The Donbass still produces over one-third of the total Soviet output of coal. In 1972 it produced 217,425,000 tons, of which 84,929,000 tons were suitable for coking. Output rose steadily up to 1971; the equivalent figures for 1965 were 205,600,000 tons and 80,600,000 tons. Production in 1972 was .02 percent below the 1971 level.[9]

The Pechora basin. Is situated in the Komi ASSR, covering an area of 120,000 sq. km., much of which is beyond the Arctic Circle. In ultimate reserves it surpasses even the Donbass. Over a third of the A + B + C_1 + C_2 reserves (estimated at 50,000 million tons in 1970) are suitable for coking, being mainly fat coal (Zh); there are also large proportions of longflame (D) and gas (G) energy coals. Reserves by grade are shown in Table 4.10.

The Vorkutskoe field has 18 mines at an average depth of 330 m., where seams averaging 1.6 m. in thickness are being worked. Coal is also being extracted from 9 mines in the Intinskoe field, and mines are being constructed at the Vorgashorskoe field.

In 1965 the Pechora basin produced 17.6 million tons of coal; by 1972 output had increased to 22,545,000 tons, of which 13,622,000 tons were suitable for coking.[10]

Legend

1 Donets (Donbass)	31 Tunguska
2 Pechora	32 Irkutsk
3 Moscow	33 Gusinoozerskoe
4 Lvov-Volyn	34 Nikolskoe
5 Dnepr	35 Bukachachinskoe
6 Novo-Dmitrovskoe	36 Chernovskoe
7 Tkvarchelskoe	37 Kharanorskoe
8 Tkibulskoe	38 Arbagarskoe
9 Akhaltsikhskoe	39 Tarbagataiskoe
10 Kizel	40 Lena
11 Chelyabinsk	41 Zyryansk
12 Serov	42 Arkagalinskoe
13 South Urals	43 Lankovskoe
14 Egorshinskoe	44 Melkovodnenskoe
15 Bulanashskoe	45 Anadyrskoe
16 North Sosva	46 Bukhta Ugolnaya
17 Karaganda	47 Taimyr
18 Samarskoe, Zavyalovskoe, Kuu-Chekinskoe	48 Korfovskoe
19 Ekibastuz	49 Krutogorovskoe
20 Maikyuben	50 South Yakut
21 Turgai (Ubagan)	51 Kivda Raichikhinsk
22 Teniz-Korzhunkul	52 Bureya
23 Alakulskoe	53 Bikinskoe
24 Lengerskoe	54 Suifun
25 Angrenskoe	55 Uglov
26 Kuznetsk (Kuzbass)	56 Suchan
27 Gorlov	57 Aleksandrovsk
28 Kansko-Achinsk	58 Uglegorsk
29 Minusinsk	59 Central Sakhalin
30 Ulukhem	60 Southern Sakhalin

MAP 4.1

Main Coal Basins and Fields

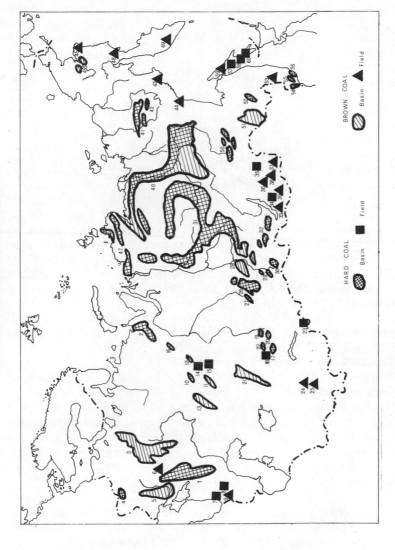

Sources: Energeticheskie resursy SSSR—Toplivno-energeticheskie resursy, Moscow 1968;
Melnikov, N. V. Toplivo-energeticheskie resursy SSSR, Moscow 1971.

141

TABLE 4.10

Pechora Balance Reserves, by Grade

Grade	Total Reserves (in thousands of millions of tons)	(in percentage)
Zh	24.17	8.3
K	4.71	1.6
OS	5.59	1.9
T + PA	15.74	5.3
G	36.13	12.4
D	152.97	52.4
B	53.14	18.1
Total	292.45	100

Source: Energeticheskie resursy SSSR: Toplivno-energeti-cheskie resursy, Moscow 1968: 80.

The Moscow basin. Covers an area of 120,000 sq. km., mainly in Tula, Ryazan, Kaluga, and Smolensk oblasts, but stretching as far north as Leningrad oblast. The coal is lignite, of low calorific value (2,690 k.cal./kg.), high sulphur content (averaging 3.7 percent), and high ash content (averaging 32 percent). It is burned locally in power stations and for railway transport.

Some 90 percent of production is in Tula oblast. In 1965 there were 109 mines working seams averaging 2.3 m. in thickness at an average depth of 45 to 60 m. Output was 40.8 million tons, of which 2.6 million was mined by opencast methods. It is calculated that by 1990 over 40 mines, with an annual capacity of 25 million tons, will have been worked out.

Production has been falling since 1958, when 47.3 million tons were extracted, because oil and natural gas are proving more economical even near the mines. In 1971 output was 36,680,000 tons, of which 25,000,000 tons were consumed in power stations. New opencast workings are being prepared, to raise extraction by this method to 6,000,000 tons a year.[11]

The Lvov-Volynsk basin. Covers an area of 1,000 sq.km. on the Polish-Ukrainian border. Its reserves and output are relatively small, but it is a valuable source of energy fuel for Belorussia and the Baltic States. Reserves are mainly gas coal (G) with 2.5 to 3.4 percent

sulphur, about 18 percent ash, and a calorific value of 5,700 k.cal./kg.
Seams are shallow (.5 to .7 m.), and mining is at depths from 300 to
500 m. Yearly output from the 18 mines is about 9 to 10 million tons.[12]

The Dnepr basin. Stretches over 100,000 sq.km. along the right bank
of the Dnepr river. Its low-calorie brown coal (1,720 k.cal./kg.) lies
at depths from 10 to 170 m., with an average seam thickness of 4 to
5 m. Output is mainly from seven open-pit workings (9.7 out of 10.8
million tons in 1965) that will be exhausted by 1980. It is hoped that
normal output will be maintained at 10 million tons by the preparation
of new open pits. The coal is made into briquettes or consumed in
power stations.[13]

The Novo-Dmitrovskoe field. Was discovered in 1961 between Izyum
and Slavyansk (Khartov oblast). Its low-quality brown coal (1,800-
2,500 k.cal./kg.), which lies at depths of 70 to 400 m., will be mined
by opencast methods and used as fuel for local power stations.[14]

The Georgian SSR. Produces less than 2.5 million tons a year. In
the Tkvarchelskoe field, $A + B + C_1$ reserves amount to 50 million
tons, mainly fat (Zh) coal with a high ash content, and there are
seven mines producing about 1 million tons per annum. Gas (G) coal
(4300 k.cal./kg.) from the four mines of the Tkibulskoe field is mixed
with coal from the above field and from the Donbass to provide coke
for the Rustavi metals plant. It has $A + B + C_1$ reserves of 230 million tons, much of which is deep-lying (800 to 1,700 m.) but in fairly
thick seams (20 to 40 m.). The brown coal from the two mines in
the Akhaltsikhskoe field is the most expensive in the USSR at 12 to 21
rubles per ton (equivalent to 35 to 50 rubles per ton of conventional
fuel), and exploitation is proving increasingly uneconomical.[15]
    There are some mines of purely local significance in Stavropol
krai and the Kabardino-Balkar ASSR, which produce under 250,000
tons a year of gas (G) coal.[16]

The Urals. Contain less than .1 percent of the total geological reserves
of coal of the USSR, only 3,500 million tons of $A + B + C_1$ reserves.
In view of the concentration of industry in the Ural economic region,
even the poorest quality local coal is being extracted, and there are
strong arguments for exploiting brown coal deposits in the Polar
regions of the Urals.

The Kizel basin. Is situated in the western slopes of the northern
Urals, stretching over 100 km. from north to south and varying in
breadth from 10 to 25 km. The five seams being worked are mainly
from .7 to 2.5 m. thick and produce fat (Zh) and gas (G) coals with

a high ash (25 percent) and sulphur (up to 6 percent) content. In 1965 the 27 mines were at average depths of 360 m., though some were as deep as 800 m. Output was 9.9 million tons, but with the exhaustion of many of the mines this is dropping to a level of 5 to 6 million tons a year.[17]

The Chelyabinsk basin. Stretches from the Techa river in the north over 200 km. to the Ui river in the south, varying in breadth from 1 to 12 km. Its brown coal reserves are approaching exhaustion. In 1965 the 27 mines and 3 open pits produced 19.7 million tons, but by the beginning of the 1970s the closure of many of the mines and two of the pits had lowered output to around 9 million tons a year. The pit at Korkino, one of the biggest in the USSR (yearly output about 4.5 million tons) is also expected to be exhausted by 1978.[18]

The Serov basin. Lies in the north of Sverdlovsk oblast and includes the Bogoslovskoe, Volchanskoe, and Veselovskoe brown coal fields, which are worked by opencast methods. Output rose to a peak of 20.1 million tons in 1965, since when it has been dropping steadily, and the pits are expected to be exhausted by 1975. The Atyusskoe field (6.8 million tons $A + B + C_1$ reserves) could be exploited to produce some .5 million tons a year, but it will still be necessary to ship in fuel for the Sverdlovsk power stations, which formerly relied on local lignite.[19]

The South Urals basin. Is situated in the Bashkir ASSR and north Orenburg oblast and has $A + B + C_1$ reserves of 1,300 million tons of poor-quality lignite; 900 million tons can be extracted by opencast methods. The Kumertau pit is producing over 6 million tons a year, and when the Tyulgan pit is brought into full operation later in the 1970s total yearly output is expected to be over 11 million tons.[20]

The Egorshinskoe and Bulanashskoe fields. In the Middle Urals, have sufficient reserves to continue producing a million tons yearly of anthracite (A) and gas (G) coal for local use until the year 2000. There are one anthracite and three gas coal mines in operation.[21]

The North Sosva basin. In the northeast Urals is the best explored area of the Sosva-Salekhard basin, which has predicted reserves of 15,500 million tons of brown coal at depths of less than 6000 m. Five fields had been explored by 1972, of which the Lyulinskoe and Otorinskoe each had some 500 million tons of reserves ($C_1 + C_2$) that were suitable for opencast working. The low quality of the coal, the considerable expenditure necessary to exploit these fields, and the availability of cheap oil and gas make it unlikely that exploitation will begin before 1980.

A strong case has been made out for greatly increasing the resources allotted by GOSPLAN and the Ministry of Geology for the exploration and utilization of the minerals in the Northern Urals, in view of the economic significance of the area. This would mean constructing a railway line from the terminus at Polunochnoe to the terminus at Labytnangi, thus allowing both the exploitation of the North Sosva basin and the direct supply of coking coal from Vorkuta. This suggestion, which was first put forward in the 1950s, was revived by the head of the Ural geology administration in 1970 and reinforced by leading fuel and energy experts in a Pravda article in August 1972; if GOSPLAN is persuaded by their logic, the railway could be built in the 1975-80 period.

An alternative, or complementary, solution to the shortage of fuel in the Urals would be to transport brown coal from the Turgai basin.[22]

The Kazakh SSR. Has some 162,400 million tons of geological coal reserves, according to the calculations of the Kazakh Ministry of Geology in 1968; some 13,800 million tons is coking coal. There are 24,300 million tons of $A + B + C_1$ reserves, most of which lie relatively near the surface and consist of high-quality coal. These factors have made the republic third in importance to the RSFSR and the Ukraine as a producer of coal; in 1970 output was over 61 million tons, and in 1975 it is expected to reach 91 million tons. An extraction rate of as much as 250 million tons is thought possible by the year 2000.[23]

The Karaganda basin. Covers an area of about 3000 sq.km. in the center of the republic. The third most important coal base in the USSR, it supplies energy and coking coal to Kazakhstan and the Urals and also sends energy coal as far as the Volga region and Central Asia. The $A + B + C_1$ reserves of the basin amount to 7,555 million tons, made up of 19 percent fat coal (Zh + KZh + GZh), 23 percent coking coal (K), 11 percent lean caking (OS), and 47 percent various energy coals. Much of the latter could also be used for coking, but its high ash content and unsuitability for dressing make it more economical to consume in power stations. There are over 40 mines in operation at an average depth of 300 m., although ten mines are worked at 400 to 470 m. The average seam is 1.8 m. thick.

Output has increased from 29.9 million tons in 1965, including 11 million tons of coking coal, to 41.7 million tons in 1972, with 17.4 million tons of coking coal. Annual production could reach as much as 50 million tons, with 29 million tons of coking coal, in the near future. In June 1972 productivity reached 81 tons per month per miner, the highest in the USSR for underground mines.

The Samarskoe field. 150 km. west of Karaganda, has reserves of
331 million tons $(A + B + C_1 + C_2)$, and the nearby Zavyalovskoe field
has reserves of 420 million tons (C_2); the coals from both are similar
in quality to those of the Karaganda basin. They are not yet in exploitation.[24]

The Kuu-Chekinskoe field. 60 km. north of Karaganda and 23 km.
east of the main railway, has $A + B + C_1$ reserves of 181 million tons,
suitable for opencast working. When full capacity is reached, annual
production will be about 3.5 million tons of energy coal.[25]

The Ekibastuz basin. Covers an area of 120 sq.km. about 130 km.
southwest of Pavlodar; it is near the main railway to Tselinograd
and is crossed by the Irtysh-Karaganda canal. It has $A + B + C_1$ reserves of 7,389 million tons at depths of less than 400 m., which can
be worked by opencast methods.

It produces a poor-quality caking coal with a high ash content
(35 to 40 percent), which cannot be economically dressed but which
serves as a valuable energy fuel. Average calorific value is 4200
k.cal./kg. A total of ten opencast workings are planned, with a potential capacity of 92 to 105 million tons a year, reaching 56 million tons
in 1975. Output in 1970 was 22.5 million tons, rising to 27 million
tons, or 40 percent of the total Kazakhstan output, in 1971. In 1972,
when output had reached 32.4 million tons, it was already supplying
coal to 17 power stations in Kazakhstan, the Urals, and Western Siberia.
Labor productivity is expected to reach 1,500 tons per man per month,
and the coal should cost an average of only .6 rubles per ton (one ruble
per ton of conventional fuel). According to a Moscow broadcast in
June 1973, two giant rotary excavators made in East Germany, with
a capacity of 3,000 tons per hour, had been installed. In January 1974,
there were reports that although output had reached 36 million tons
in 1973, this was 2.5 million tons below target; poor organization,
delays in the transport system, and inadequate equipment were blamed.[26]

The Maikyuben basin. Is situated 150 km. southwest of Pavlodar near
the railway to Tselinograd and covers an area of 1,400 sq.km. It has
$A + B + C_1$ brown coal reserves of 1,760 million tons and C_2 reserves
of 3,180 million tons. The Shoptykolskoe field could support an opencast with a yearly capacity of 20 million tons. This basin will be developed in association with the Ekibastuz complex, on the basis of
which the North Kazakhstan energy complex is planned, with an aggregate capacity of almost 16,000 MW. Electricity will be transmitted
to the Central region by a DC line of 1500 kV. and to Chelyabinsk by
AC line of 760 kV.[27]

The Turgai basin. Earlier known as the Ubagan basin, contains about 20 brown coal fields with total geological reserves of 61,900 million tons, stretching along the Tselinograd-Magnitogorsk railway in the Kustanai and Tselinograd oblasts. Many seams are 40 to 70 m. thick and suitable for opencast working. The Kushmurunskoe and Priozernoe fields, with 2,783 million tons and 391 million tons of $A + B + C_1$ reserves, respectively, are the most promising; the Eginsaiskoe and Orlovskoe fields, with 1,097 and 992 million tons of $A + B + C_1$ reserves, could also be brought into production. Yearly output could be as much as 100 to 120 million tons, sufficient to provide fuel for local and Ural power stations with a total capacity of over 12,000 MW. The Turgai Basin is only 200 to 500 km. from major Ural consumers, compared with distances of 1,200 km. for the Ekibastuz and 2,500 km. for the Kansko-Achinsk basins. In spite of strong arguments for its immediate exploitation, the Turgai basin was still being held in reserve in 1973.[28]

The Teniz-Korzhunkul basin. Is situated 160 km. northwest of Tselinograd and has hard coal $A + B + C_1$ reserves of 315 million tons. Although close to the main railway line, it is too near Ekibastuz for exploitation to be economically viable within the near future.

Other Kazakh coal fields, such as the Alakulskoe, near Alma-Ata, and the Lengerskoe, near Chimkent, are only of local significance.[29]

The republics of Soviet Central Asia. Have total balance geological coal reserves of 38,300 million tons, almost half of which are situated at depths of less than 600 m. Transporting coal to points of consumption often presents great difficulties, however, and many comparatively large fields remain unexploited. The distribution of the explored balance reserves as of 1966 is shown in Table 4.11. Coal fields are mainly of local significance, although some coal is transported over short distances between republics. Among the largest producers is the Angrenskoe brown coal field in the Uzbek SSR, some 110 km. east of Tashkent, where opencast methods are used to extract some 3 million tons a year. Since much of the coal output is from the Fergana Valley, with its confused maze of boundaries, statistics for each republic are of limited value. In 1970 the coal industries of the Uzbek, Kirgiz, and Tadzhik republics were merged in one production association, the "Sredaz-ugol" combine, and in that year the three republics together produced only 8.3 million tons; this represents a small but steady annual drop since 1965, when the 16 mines and 3 open pits in the Central Asian republics produced 9.6 million tons. The increasing availability of natural gas makes it unlikely that the coal industry in Central Asia will need to expand coal output to any great extent in

TABLE 4.11

Explored Balance Coal Reserves in Central
Asia $(A+B+C_1+C_2)$
(in millions of tons)

Republic or Field	Hard	Brown
Kirgiz SSR	1,419	2,096
Kok-Yangak	157	—
Naryn	88	—
Kyzyl-Kiya	87	—
Sulyukta	—	217
Shurab III	—	336
Kara Kiche	—	414
Kok-Mainak	—	143
Minkushi	—	730
Dzhergalan	38	—
Uzgen	960	—
Uzbek SSR	143	2,028
Angren	—	2,028
Shargun	42	—
Tadzhik SSR	1,031	159
Shurab I, II	—	159
Fan Yagnob	846	—
Total for Central Asia	2,593	4,286

Source: Energeticheskie resursy SSSR: Toplivno-energeticheskie
resursy, Moscow 1968: 103.

the near future, in spite of the relatively high quality of the coal being
extracted, in which ash content varies from .4 to 2.3 percent and
calorific value is between 3,400 and 5,100 k.cal./kg. In 1971 as much
as 46 percent of the 8.4 million tons mined was by opencast methods,
which are to account for the expected increase to the 1975 target of
9.8 million tons.[30]

In Western Siberia. The total balance geological reserves were estimated in 1956 at 849,000 million tons, of which 83,100 million tons
were proven to categories $A + B + C_1$ by 1966.

The Kuznetsk basin (Kuzbass). Covers an area of 26,000 sq.km. in the Kemerovo oblast. It is the fourth largest coal basin in the USSR, with reserves greater than those of the Donets, Pechora, and Karaganda basins combined. It not only has over 10 percent of the total Soviet balance geological reserves (804,200 million tons) but it is estimated that almost a third of this is coking coal. The basin is comparatively well explored, with $A + B + C_1$ reserves of 52,510 million tons as of 1966, but as yet the central and eastern areas of the basin are little developed.

Coal seams fluctuate significantly in thickness, mainly from 1 to 12 m., with an aggregate thickness of up to 99 m.; in some parts around the Kemerovo and Prokopev-Kiselev regions the seams are steeply inclined, while those in the north tend to be more gently sloping. Most mines are not at great depths, only approaching 500 m. in a few regions; the average is 185 m. The average thickness of seams being worked varies according to region from 1 to 3 m. The presence of gas is a serious problem in certain seams, which is an additional factor favoring strip mining. Unfortunately, almost two-thirds of the total reserves are at depths greater than 600 m., and only 17 percent at less than 300 m.

The classification of the coal being produced is as follows: 8 percent gas coking coal (G); 12 percent fat coal (Zh+PZh+GZh); 26 percent coking coal (K+KZh); 20 percent caking coal (OS+SS); 19 percent lean coal (T), and 15 percent of various energy coals. Ash content is low, mostly 6 to 15 percent, and the coals also contain little sulphur, averaging from .4 to .6 percent, while calorific value is high (5,900 to 6,900 k.cal./kg.) The total balance reserves are given by grade in Table 4.12.

According to figures published in 1971, some 10,368 million tons of the Kuznetsk coal reserves $(A + B + C_1)$ are accessible to strip mining, and it was calculated that extraction by this method could be raised to 95 million tons a year, of which almost 30 percent would be suitable for coking. The first open pits began production in 1949, and by 1970 there were sixteen open strip mines with another 3 under construction. In 1972 some 32 million tons were extracted by strip mining. The Kedrovo, Tomusinsk, and Krasnogorsk open pits each produce over 3 million tons annually.

Most of the output of coal in the Kuznetsk basin, however, is from about 100 underground mines. Total output has increased from 97 million tons in 1965, including 38 million tons of coking coal, to 119.2 million in 1972, of which 50.1 million tons was suitable for coking, and was expected to reach 135 million tons in 1975. It is estimated that in future years as much as 200 million tons, including 75 million tons of coking coal, could be extracted annually.[31]

TABLE 4.12

Kuznetsk Total Balance Reserves, by Grade

Grade	Reserves (in thousands of millions of tons)	Percentage
B + DB	45.2	5.6
D	61.1	7.6
G	271.5	33.9
Zh	56.3	7.0
K	37.8	4.7
OS	88.6	11.0
T + PA	243.7	30.4
Total	804.2	100

Source: Energeticheskie resursy SSSR: Toplivno-energeticheskie resursy, Moscow 1968: 108.

The Gorlov basin. 100 km. southeast of Novosibirsk has balance geological reserves of 15,160 million tons of high-quality anthracite (ash content is under 7 percent, sulphur content .3 percent). In 1965 the two mines in the Listvyanskoe field produced 216,000 tons.[32]

Eastern Siberia. Contains over 45 percent of the geological coal reserves of the USSR, but is as yet comparatively poorly explored, especially in the north, where the Tunguska and Taimyr basins are situated. The best explored areas are in the south, where transport systems are already in existence.

The Kansko-Achinsk basin. In the Krasnoyarsk krai, covers a vast area 700 km. long and 50 to 300 km. wide and is the third largest basin in the USSR, with over 15 percent of the total geological reserves. Its $A + B + C_1 + C_2$ reserves were stated in 1970 to be 610,000 million tons. It is traversed by the Trans-Siberian railway and by the main lines from Achinsk to Abakan and Abakan to Taishet. In 1966 the $A + B + C_1$ reserves were reckoned at 68,100 million tons of good-quality brown coal, with an ash content of between 7 and 20 percent, sulphur content of .8 to 1 percent, and calorific value of 2,800 to 3,600. The main seam is 13 to 50 m. thick, mainly horizontal and at depths less than 100 m., which makes it ideal for strip mining. As much as 80,000 million tons may be mined by this method.

The Nazarovo and Irsha-Borodino open pits, which began exploitation in 1950, were each producing over 10 million tons a year by the 1970s at a cost per ton of only 1.15 rubles. Production per man per month was 564 and 642 tons, respectively. Yearly capacity at the Nazarovo pit is planned to reach 16 million tons by 1977, while at the Irsha-Borodino pit annual output should reach 45 million tons by the 1980s. The Berezovskoe pit is being prepared for exploitation, and is expected to produce about 20 million tons a year by 1980, increasing eventually to as much as 60 million tons a year. An equal capacity is planned for the Itatskii pit. Total output at the Kansko-Achinsk basin is planned to reach 33 million tons in 1975 and about 50 million tons in 1980, over 90 percent by opencast methods.

Some idea of the possible advantage to the Soviet economy of increasing the consumption of coal from the Kansko-Achinsk basin can be gained from a comparison of fuel costs. In 1970 the cost of extracting a ton of coal at Kuzbass was 14.5 rubles, at Ekibastuz 4.3 rubles, but at Kansko-Achinsk only 2.5 rubles. In general the cheapest fuel is natural gas, but in terms of conventional fuel, coal from Kansko-Achinsk costs 1.8 rubles per ton compared to 2.2 rubles per ton for Tyumen gas and 5.4 rubles per ton for Turkmen gas.

These costs obviously vary greatly when transport and manner of consumption are considered, but there is nonetheless a strong argument for constantly expanding production at the Kansko-Achinsk basin. It has been calculated that yearly output could approach 600 million tons by the year 2000, when average output per man per month would be as much as 2,500 tons and cost per ton as low as 50 to 70 kopeks.

Electric power produced from this cheap fuel could cost less than .18 kopeks per kWh, under half the present cost in West Siberian and Urals thermal power stations. It is intended to construct an energy complex based on coal from the western Kansko-Achinsk fields that could have an aggregate capacity of 34,000 MW by the 1980s, rising to an eventual 45,000 to 50,000 MW. This would make it possible to supply electricity to the Urals by high-tension line at a cost of only .16 kopeks per kWh, when electricity at the point of origin, Itatskii, costs .12 kopeks per kWh.

In addition to brown coal, the Kansko-Achinsk basin has some hard coal reserves at the Sayano-Partizanskoe field. It is gas (G) medium caking coal with a low ash and sulphur content. Extraction is possible only by underground mine.[33]

The Minusinsk basin. Is situated in the Khakas autonomous oblast, around the junction of the Abakan and Enisei rivers. It is served by the Novokuznetsk-Abakan-Taishet railway. It has 36,300 million tons of balance geological reserves in seams ranging from 1 to 20 m. in thickness, consisting of longflame (D) and gas (G) coal with moisture content of 13 to 15 percent, ash content of 10 to 15 percent, and sulphur content of .5 to 1 percent. Calorific value ranges from 4,900 to 5,580 k.cal./kg. There are four coal fields: the Beiskoe and Askizskoe, in reserve; the Chernogorskoe, brought into exploitation in 1957; and the Izykhskoe, brought into exploitation in 1961. Output is from eight underground mines and two opencast pits, with a total annual capacity in 1965 of 3.6 million tons.

Production by strip mining is expected to rise from the 2.3 million tons mined in 1971 to 8 million tons in 1980. The Beiskoe field has a potential output from five open pits of 4.6 million tons a year, raising the total yearly strip mining capacity of the basin to 54 million tons.[34]

The Ulukhem basin. Occupies an area of 2,500 sq.km. in the Tuva ASSR and has balance geological reserves of 10,000 million tons in seams ranging from 2 to 10 m. in thickness. The coals, which consist equally of gas (G) and fat caking (Zh) classes, have an ash content of 6 to 25 percent, sulphur content of .4 to .9 percent, and a calorific value of 6,500 to 7,500 k.cal./kg. Coal has been extracted since 1962 near Kyzyl, the annual output of under 200,000 tons being sufficient to satisfy local needs.[35]

The Tunguska basin. With almost 20 percent of Soviet geological coal reserves, is second only to the Lena basin in size. It stretches over almost one million sq.km. in Krasnoyarsk krai, between the Enisei and Lena rivers. It is very difficult to reach and has been little explored. Only in the northwestern part, near Norilsk, is it being exploited.

There are seven mines and two opencast workings, producing under 4 million tons a year. Coal in some explored areas is longflame (D) with 6 to 10 percent ash content, .3 to .6 percent sulphur content, and calorific value of 6,500 to 7,500 k.cal./kg. The Norilskoe and Kaierkanskoe fields are already being exploited, while the Imangdinskoe and Kokuiskoe will not be brought into production until local requirements for energy coal are considerably increased. The main consumers at present are the Norilsk non-ferrous-metals combine and the Dudinka river fleet. Since 1971 natural gas from Messoyakha has been the main source of energy for Norilsk, and coal output has been dropping.[36]

The Irkutsk basin. Is traversed by the Trans-Siberian railway and Angara river, which provide a useful outlet to the industrial regions of the Irkutsk oblast. It has balance geological reserves of 67,300 million tons and is the best explored of the East Siberian basins. The most important field is the Cheremkhovskoe, with seven open pits and four underground mines, which in 1965 produced 12 million tons

152

and 3 million tons respectively of gas (G) coal with ash content of 16 to 25 percent, average sulphur content 1.4 to 1.6 percent, and calorific value 4,850 to 5,420 k.cal./kg. There are two open pits producing brown coal at the Azeya field, with a total yearly capacity of 9.5 million tons, ash content 12 to 20 percent, and sulphur content .3 to .7 percent. In general, the brown coal of the Irkutsk basin has a high ash content and low sulphur content, while the hard coal (G+D) has a low ash content but high sulphur content, up to 8 percent. It is more successfully used as a power fuel and chemical raw material than as coke.

Some 65 percent of the explored reserves can be mined by opencast methods. In 1968 a total of 21.3 million tons were extracted, of which 2.2 million tons were brown coal. Over 83 percent was from open pits. Output is expected to continue to increase throughout the 1970s.[37]

The Transbaikal region. Has many small coal fields, with total balance geological reserves of 7,200 million tons. By 1970 twelve of the fields had been explored and prepared for exploitation. In the Buryat ASSR the most important field is the Gusinoozerskoe, which produces about one million tons of brown coal a year for local power stations. The Nikolskoe field has sufficient reserves to allow the extraction of up to 5 million tons annually by opencast methods. The coal here is of the longflame gas (DG) type, ash content 15.5 to 17.5 percent, sulphur content under .5 percent, and calorific value 5,300 to 5,600 k.cal./kg. The Akhalikskoe field also has substantial reserves suitable for strip mining. In Chita oblast the largest producer is the Bukachachinskoe field, which has one mine with an annual output of one million tons of good-quality gas (G) coal. The Chernovskoe field is approaching exhaustion, but still produces over one million tons of high-quality brown coal a year. The greatest reserves, 1,084 million tons of $A + B + C_1$, are in the Kharanorskoe brown coal field, which could produce as much as 15 million tons annually by opencast methods. The Arbagarskoe and Tarbagataiskoe brown coal fields have also substantial reserves, although present output is low.

Total output for 1968 in the Transbaikal region was about 5 million tons, of which almost half was by strip mining. Most of the fields that are being exploited have access to the Trans-Siberian railway.[38]

There are vast coal reserves in the Northeast of the USSR, but because of the harsh climate and lack of transport facilities, they have been but little explored and have not as yet been exploited to any extent.

The Lena basin. Which has over 31 percent of the total balance geological reserves, is the largest coal basin in the USSR. It stretches

over 400,000 sq.km. in the northern and central parts of the Yakut
ASSR, its contours marked by the course of the Lena river and its
tributaries, the Aldan and Vilyui. By 1970 the A + B + C_1 reserves
amounted to 830 million tons, of which over half could be extracted
by strip mining. The coal consists of 57 percent brown coal and 28
percent longflame (D). The brown coal has an ash content ranging
from 6 to 15 percent and sulphur content from .2 to .5 percent.

Output is insignificant when compared to reserves. In 1965
some 344,000 tons of longflame (D) coal were extracted at the Sangarskoe
field, 330 km. downriver from Yakutsk; 353,000 tons at the Dzhebariki-
Khayaskoe field, on the river Aldan; 337,000 tons of brown coal at the
Kangalasskoe field, 45 km. north of Yakutsk; and 108,000 tons of brown
coal at the Soginskoe field, near Tiksi Bay. By 1968 output at the
Kangalasskoe field had dropped by 120,000 tons because of the conversion of the Yakutsk power station to natural gas.[39]

Some idea of the variety of problems involved in exploiting the
coal resources of the inaccessible areas of the USSR can be gained
from the example of the Sangarskoe field. The average surface temperature is below -10°C, and even in summer there are strong winds.
Long-distance transport is mainly by river in summer and by air in
winter, weather permitting. There have been several explosions and
fires in the mines due to methane gas and coal dust; such an explosion
occurred as recently as 1966. In winter the temperature even deep
in the mines can drop as low as -9°C, in spite of attempts in 1969 to
heat the air entering the mines. Coal is actually supplied not only
to enterprises on the Lena river but also to those on the Yana, Indigirka,
and Anabar rivers. As this involves transferring the coal from river
barges to oceangoing vessels and back to river boats, it is not surprising that coal that costs 12.85 rubles per ton at the pit head can
cost as much as 108 rubles per ton at Deputatskii, on the Uyandina
tributary of the Indigirka river.[40]

The Zyryansk basin. Is situated in the east of the Yakut ASSR on the
upper reaches of the Kolyma river. In 1970 the explored deposits
had A + B + C_1 reserves of 100 million tons, of which 17 percent was
suitable for strip mining. In 1969 opencast workings at the Erozionnoe
field produced 199,000 tons of coking (K+Zh) coal. Coal is shipped
from the Zyryansk basin as far as Pevek and Bilibino.[41]

Magadan oblast. The main coal field is the Arkagalinskoe, on
the left bank of the upper Kolyma river. Though not fully explored
in 1970 the A + B + C_1 reserves amounted to 20 million tons of gas
(G) and longflame (D) coal. It produces about one million tons annually
to meet the energy requirements of the oblast. Some coal is transported to the gold mines on the upper Indigirka river. The Lankovskoe

brown coal field, 70 km. from Magadan, has $A + B + C_1$ reserves of
138 million tons. It is intended to supply 1.5 million tons a year to
the Magadan heat and power station. The Melkovodnenskoe field on
the Okhotsk coast has $A + B + C_1$ reserves of 160 million tons of brown
coal. The Anadyrskoe brown coal field, with $A + B + C_1$ reserves of
107 million tons, produces about .1 million tons a year. The Bukhta
Ugolnaya fields near Beringovskii have $A + B + C_1$ reserves of 386
million tons and produce about .2 million tons of hard coal a year.[42]

Kamchatka oblast. The Korfovskoe field supplies the local fishing
industry on the northeast coast with some 10,000 tons of brown coal
a year. Exploitation of the Krutogorovskoe field near Sobolevo is
not yet necessary. Each field has $A + B + C_1$ reserves of about 28
million tons.[43]

     In the more southern area of the Soviet Far East there are several
important coal basins.

The South Yakut basin. Is situated in the upper reaches of the Aldan
river, about 400 km. north of the Trans-Siberian railway, to which
it is connected by the Magadan-Never highway. In 1966 the $A + B + C_1$
reserves in the two fields that had been explored amounted to 2,049
million tons. The Chulmakanskoe field has mainly fat (Zh) coal with
low sulphur content, and ash content ranging from 13 to 26 percent.
Mining began in 1966 by the room-and-pillar method; annual output
in the first years was about 100,000 tons. The Neryungrinskoe field
is worked by strip mining; it produced 122,000 tons of coking (K) coal
in 1968 and has a present planned capacity of 375,000 tons. It supplies
coal to the Chulman power station and to the gold and mica enterprises
of Aldan. Coal from the two fields can be mined to form high-quality
coke.[44]

The Kivda-Raichikhinsk. Coal region is situated between the Raichikha
and the Kivda tributaries of the Amur river. In 1966 the $A + B + C_1$
reserves amounted to 377 million tons of brown coal with a low sulphur
and ash content. Yearly output from the four opencast workings is
about 12 million tons. In 1963 a new field was discovered near Svobod-
nyi with geological balance reserves of 1,691 million tons. The
Svobodnoe field has a planned annual output of 15 million tons of brown
coal. The Ogodzhinskoe hard coal field, in the northeast of Amur
oblast, produces about 80,000 tons a year for the Ogodzha heat and
power plant.[45]

The Bureya basin. Occupies an area of 6,000 sq.km. in Khabarovsk
krai. The best explored part is the Urgalskoe field, with $A + B + C_1$
reserves in 1966 of 506 million tons. The coal is gas (G) grade, low

in sulphur (.6 percent), with a high ash content (mostly about 31 to 33 percent), and calorific value of about 5,000 k.cal./kg. It is difficult to dress, but after complex processing may be used for coking. In 1965 some 1.2 million tons were produced, and because of the development of local industry an annual output of 5 million tons is expected in the 1970s.[46]

The Bikinskoe field. Is situated some 250 km. southwest of Khabarovsk. All of the 729 million tons of A + B + C_1 brown coal reserves can be extracted by strip mining. An opencast working with a potential yearly output of 8 million tons was being prepared in 1972.[47]

The Suchan basin. 150 km. east of Vladivostok, had A + B + C_1 reserves of 202 million tons in 1966. Output from nine mines is being maintained at about 2 million tons a year of energy coal. The Chinese name Suchan was changed to the Russian name Partizansk in 1973.[48]

The Uglov basin. North of Vladivostok, had A + B + C_1 reserves of 248 million tons at the Artemovskoe field in 1966 and 57 million tons at the Tavrichanskoe field; about 3 million tons of brown coal is extracted annually from its six mines.[49]

The Suifun basin. Consists of several small fields northeast of Vladivostok that produce about one million tons a year of longflame (D) and lean (T) coal.

The Rettikhovskoe field. With A + B + C_1 reserves of 38 million tons, produces about 1.2 million tons of brown coal annually by opencast methods.

The Chikhezskoe field. With A + B + C_1 reserves of 353 million tons of brown coal, is planned to support a yearly extraction of 3 million tons by opencast methods.[50]

Sakhalin Island. There are many coal fields, which are generally grouped into four main areas: Aleksandrovsk, with 206 million tons of A + B + C_1 reserves of hard coal; Uglegorsk, with 552 million tons of hard coal and 90 million tons of brown coal; Central, with 340 million tons of brown coal; and Southern, with 292 million tons of hard coal and 593 million tons of brown coal. In 1965 the sixteen Sakhalin mines produced 4.2 million tons; by 1970 an additional 1.2 million tons was being extracted from the two open pits. Output at the Vakhrushevskoe field, 40 km. north of Makarov, is planned to reach 1.5 million tons by strip mining.[51]

Production Statistics

By the beginning of the 1970s the total output of coal by the
nations of the world was almost 3,000 million tons a year. The USSR,
which in 1971 extracted some 641 million tons of coal, is the major
producing nation, although a large proportion of its output is made
up of brown coal. The United States still produces more bituminous
coal than the USSR. In view of the wide variations in the quality of
coals from different fields and the diverse national methods of classi-
fying and grading these coals, it is difficult to compile statistics on
the comparative distribution of world coal extraction. Nonetheless,
a fairly accurate picture of trends in world production since 1950 is
given in Table 4.13. Precise information on China is not available,
but yearly output is probably about 400 million tons.

A considerable proportion of the world's output of coal is therefore attributable to Eastern Europe and the USSR. By far the largest
individual producers are the USSR and the United States; their annual
output since 1960 is compared in Table 4.14.

Accepting the Soviet conversion factor of one ton of brown coal
equivalent in energy content to only .4 tons of hard coal, it will be
seen that in 1960 the USSR surpassed the United States in production,
only to drop behind as output in the United States again began to increase. Estimates for the year 2000, however, would suggest that
the United States will not retain its lead permanently; possible extraction rates for that year are 1,100 million tons in the United States,
compared with 1,700 million tons in the USSR. Such predictions are
supported by the coal reserves in each country. (See Figure 4.3.)[52]

Within the USSR the main producing republics are the RSFSR,
with over half the annual output; the Ukraine, with about one-third;
and Kazakhstan, with about one-tenth. The regional distribution of
the Soviet coal-extracting industry has been discussed above. (See
especially Table 4.3.) The actual output by the main basins and fields
since 1960 is given in Table 4.15 and Map 4.2.

Some 90 percent of the total Soviet coal reserves lie beyond
the Urals, but as yet the distribution of production corresponds less
to the locations of reserves than to the areas of greatest demand.
The share of the basins and fields in Asian USSR is nonetheless steadily
growing, and the trend is expected to continue in future years. Since
1955 the rate of increase in coal extraction of Asian USSR has been
double that of European USSR; this is largely due to the extensive
application of the latest strip mining technology in the brown coal
deposits that are now being exploited beyond the Urals.

In 1940 as much as 75 percent of Soviet coking coal was produced
in the Donbass; by 1970 output was more evenly distributed, as is
shown in Table 4.16. The proportion attributable to the Donbass has

TABLE 4.13

Production of Hard and Brown Coal by Major Producing Nations, 1950-70
(in millions of metric tons)

Nation	Hard Coal				Brown Coal			
	1950	1960	1965	1970	1950	1960	1965	1970
World	1,436	1,962	2,046	2,126	381	635	740	792
USSR	185	356	398	433	76	134	147	145
United States	505	392	475	542	3	2	3	5
West Germany	126	143	135	111	76	96	102	108
East Germany	2	3	2	1	137	225	251	261
United Kingdom	220	197	191	145	—	—	—	—
Poland	73	104	119	140	5	9	23	33
Czechoslovakia	18	26	28	28	28	58	73	82

Source: UN Statistical Yearbooks 1950-70, New York 1951-71.

158

TABLE 4.14

Coal Production in the USSR and United States, 1960-71
(in millions of metric tons)

Year	USSR		United States	
	Hard	Brown	Hard	Brown
1960	355.9	134.2	391.5	2.5
1961	355.8	128.6	378.7	2.7
1962	363.4	130.1	395.5	2.8
1963	369.3	135.7	430.5	2.5
1964	381.3	142.9	454.7	2.7
1965	397.6	147.4	475.3	2.8
1966	406.6	144.2	492.5	3.5
1967	414.1	141.4	508.4	4.1
1968	416.2	135.9	500.7	4.4
1969	425.8	137.7	513.4	4.5
1970	432.7	144.7	541.6	5.4
1971	441.4	150.1	503.1	5.8

Sources: UN Statistical Yearbook 1970, New York 1971: 165;
UN Statistical Yearbook 1971, New York 1972: 164-166. UN Statistical
Yearbook 1972, New York 1973: 178-179.

been dropping steadily, being slightly over 50 percent in 1971, while
the share of the Kuzbass has risen to about 30 percent. Production
costs and the quantity and quality of reserves are much more favorable
in the latter, but the expense of transporting Kuznetsk coal over long
distances to the main areas of consumption has tended to retard its
development to some extent. The high ash content of Karaganda coal
necessitates the addition of Kuznetsk coke to make it suitable for the
metals industry. The high production costs and remote location of
the Pechora Basin have restrained the growth in extraction rates for
Pechora coke.

Methods of Production

In 1972 as much as 71 percent of Soviet coal was still being
extracted from underground mines, and much effort has been devoted
to improving the standard of technology available for this sector of
the industry. The process of cutting and removing coal from the

FIGURE 4.3

Coal Production in USSR and United States, 1960–1972

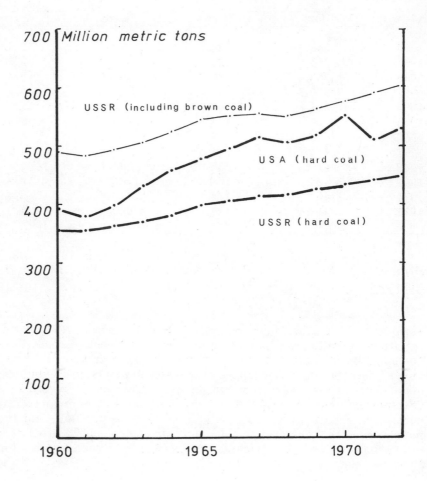

Source: Table 4.14.

TABLE 4.15

Coal Output in USSR, by Basin and Field, 1960–72
(in millions of tons)

Basin or Field	1960	1965	1967	1969	1970	1971	1972
Donets	188.2	205.9	210.2	213.9	216.1	217.5	217.4
Pechora	17.6	18.1	20.0	21.2	21.5	22.0	22.5
Moscow	42.8	40.8	37.9	35.1	36.2	36.7	36.7
Lvov–Volyn	3.9	9.8	11.1	12.1	NA	NA	NA
Georgia	2.9	2.6	2.4	2.3	2.3	2.2	NA
Urals	61.2	61.6	57.0	52.7	NA	NA	NA
Karaganda	25.8	30.9	33.3	36.0	38.4	39.8	41.7
Ekibastuz	6.0	14.3	16.9	21.0	22.5	27.0	32.4
Kuznetsk	84.1	96.9	104.9	107.7	110.5	115.5	119.2
Central Asian fields	7.8	9.1	8.9	8.6	8.3	8.0	NA
Eastern Siberian fields	36.9	48.3	52.7	48.4	56.8	NA	62.6
Far East fields	21.9	27.0	27.5	28.7	NA	NA	NA

NA: Not available.

Sources: Melnikov, N. V., Mineralnoe Toplivo, Moscow 1971: 194; Narodnoe Khozyaistvo SSSR v 1970 godu, Moscow 1971: 188; Ugol no. 4, April 1973: 72; Ekonomicheskaya gazeta no. 6, February 1973: 2.

MAP 4.2

Coal Output by Basin

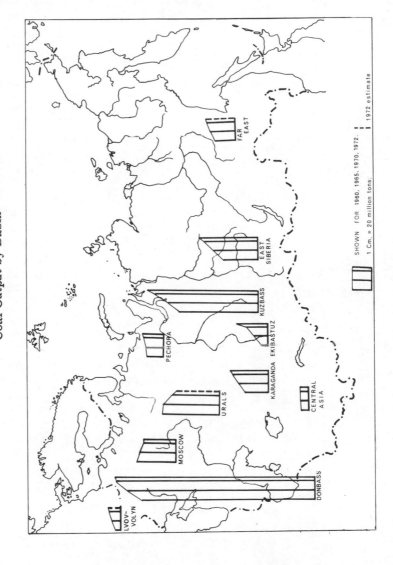

SHOWN FOR 1960, 1965, 1970, 1972. ---- 1972 estimate

1 Cm. = 20 million tons.

LVOV-VOLYN

DONBASS

MOSCOW

URALS

PECHORA

KUZBASS

EAST SIBERIA

FAR EAST

KARAGANDA EKIBASTUZ

CENTRAL ASIA

Source: Table 4.15.

TABLE 4.16

Production of Coking Coal in USSR, by
Republic and Basin, 1965-72
(in millions of tons)

	1965	1970	1971	1972
USSR	139.0	164.8	169.2	170.0
Republic				
RSFSR	48.9	65.4	68.6	69.6
Ukraine	77.0	80.7	81.9	81.2
Kazakhstan	11.0	17.0	16.9	17.4
Georgia	2.1	1.8	1.8	1.8
Basin				
Donets	80.6	84.3	85.4	84.9
Kuznetsk	33.0	46.9	49.8	50.1
Karaganda	11.0	17.0	16.9	17.4
Pechora	4.7	12.7	13.2	13.6
Others	4.7	3.9	3.9	4.0

Sources: Narodnoe khozyaistvo SSSR v 1970 godu, Moscow 1971: 188; Ugol no. 4, April 1971: 70; Ugol no. 4, April 1972: 69; Ugol no. 4, April 1973: 72.

face has been almost completely mechanized. Progress in this sphere is shown in Table 4.17.

The use of large combines for cutting and loading coal onto conveyors has increased greatly since 1940, when there were only 24 fairly elementary types in the whole of the USSR. By 1950 the number of combines had risen to 688, and in 1970 there were 4,154 advanced combines in Soviet mines. In 1970 the USSR had some 840 mines, in which among other major items of equipment were over 50,600 conveyors, 13,800 electric locomotives, 600 heavy coal cutters, and 4,500 loaders. The numbers of the last two items in use have dropped considerably since the introduction of combines; in 1960, for example, there had been 3,800 heavy coal cutters and 6,600 loaders.[53]

By 1972 some 630 coal faces out of a total of 5,500 had been completely mechanized, using various combinations of timbering machines, combines, and conveyors. Capital expenditure is usually recovered in less than 18 months because of savings in labor costs.

TABLE 4.17

Mechanization of Coal Mining Processes in USSR, 1940-70

Indicator	1940	1950	1960	1970
Coal cutting	94.8	98.7	99.2	99.9
Loading at coal-face in pitching seams	0.1	15.7	48.8	84.9
Conveying coal from face	90.4	99.1	99.9	100.0
Coal and rock haulage	75.2	97.1	99.9	100.0
Loading of coal and rocks while driving main tunnels	—	24.6	67.0	86.8
Loading coal into railway wagons	86.5	99.4	100.0	100.0

Source: Narodnoe khozyaistvo SSSR v 1970 godu, Moscow 1971: 189.

Increased automation and remote control have also improved safety for miners. Among the most original developments in this field have been the successful experiments carried out in the USSR with methane-consuming bacteria to reduce explosions in the mines.

Since 1960 many small, uneconomical mines have been closed, the total number of mines dropping from 1,080 to 842 by the beginning of 1970. In the same period, however, average annual output increased from 360,000 tons to 514,000 tons per mine, and labor productivity in the mines rose from 35.7 to 44.3 tons per man per month. In 1972 the average output per man per month reached 51 tons in underground mines.

In 1973 the "Raspadskaya" mine in the Kuzbass was being constructed, with a capacity of 6 million tons a year, and the "Vorga-shorskaya" mine in Vorkuta, with a capacity of 4.5 million tons. Labor productivity in these two mines was expected to reach 201 and 214.5 tons per man per month, respectively.

Progress has been restrained to some extent by the gradual exhaustion of the more accessible seams and the subsequent need to mine deeper, thinner, and more steeply pitched seams. The average depth increased from 223 m. in 1958 to 323 m. in 1965, and in the same period the proportion of gas-free mines dropped from 39 percent to 27 percent. Such natural difficulties have not been the only problems, however, and there have been many criticisms leveled at the organization and management of coal mining enterprises. Equipment has been badly maintained and inefficiently used, and there have been many cases of underemployment among servicing personnel because of the "hoarding" of skilled labor by some managers.[54]

In August 1970 the Minister of the Coal Industry, Boris Fedorovich Bratchenko, claimed that the rate of mechanization and the improvement in labor productivity was not satisfactory. He blamed the heavy engineering industry for poor-quality equipment and for insufficient numbers of new machines, but also maintained that existing technology was not always fully utilized. Research and planning institutes were not meeting the demands for new methods and technology. He demanded more efficient management and the streamlining of the administrative structure.[55]

By September 1972, however, he was able to report that administration had been improved by the elimination of the ninety trusts ("tresty") and of the Kirghiz and Uzbek mine groups ("kombinaty"). The simplified structure was achieved by bringing mining enterprises ("predpriyatie") directly under 38 administrative groups ("kombinaty"), 23 of which were responsible to the USSR Coal Ministry and 15 to the Ukraine Coal Ministry. Some 22 of the administrative groups with automated management systems were being linked with a central computer, which could thus supply immediate information on the running of the main coal basins. The system incorporates 550 mines and opencasts, which together produce about three-quarters of total output.[56]

Cutting coal by hydraulic methods has been increased in recent years because of the advantages in reducing the labor force and lowering the accident rates. Experiments in cutting and transporting coal using high-pressure water jets began in the 1930s at the Kizel basin. The greatest developments have been at the Kuznetsk, Donets, and Karaganda basins. Productivity of 127 tons per man per month has been attained at one of the Kuzbass mines; cost per ton (1965) was 4.71 rubles, compared with the average in neighboring mines, using normal methods, of 7.68 rubles. In the USSR as a whole, however, the average saving from using hydraulic methods was only 6 percent a ton, with average productivity around 60 tons per man per month.

By the beginning of the 1970s the total capacity of "hydro-mines" was over 13 million tons a year.[57]

In spite of the mechanization of the processes of coal extraction over the past decades, there has not always been a significant drop in the number of miners involved in hard physical labor. In the Ukraine, for example, the proportion of miners accomplishing their tasks using hand tools dropped a mere 4.3 percent between 1959 and 1969, from

76.1 percent to 71.8 percent. Those involved in jobs defined as "hard physical labor" dropped in the same period from 67.4 percent to 58.4 percent.

The large volume of work still being done by hand in the USSR can be explained to some extent by the poor technical quality of some of the mining equipment being produced. According to an article published in May 1972, the number of men using automated machinery had increased by 4.1 percent over the past decade, while the number involved in repairing and adjusting the new equipment increased by 4.8 percent.

Action is being taken to improve the situation, and it is planned to reduce the amount of work done by hand in the Soviet coal industry by about 4 percent in the 1971-75 period, thus releasing some 16,000 men for other tasks.

The necessity for improving technology rapidly in the coal industry has been recognized at the highest level. The CPSU Central Committee issued a statement in April 1973 claiming that standards in the industry were still not matching demands. Many labor-consuming processes had not yet been mechanized, and many coal-faces were still equipped with out-of-date machinery. Design and construction organizations and research institutes under the Ministry of the Coal Industry and the Ministry of Heavy Engineering had not yet developed effective techniques for raising coal under difficult conditions. The ministers concerned, Bratchenko and Zhigalin, were held personally responsible for eliminating the many shortcomings that had become apparent. No immediate improvements were effected, however, and severe criticisms of these ministries at the highest level continued in 1974.[58]

Possibly the most important single development in the Soviet coal industry has been the rapid expansion of opencast output since the war. (See Table 4.18.)

Four basic methods are used in Soviet opencast workings. The most economical is by using excavators alone to remove the over-burden into areas already worked out ("bestransportnaya sistema"). Self-propelling draglines are being constructed in ever-increasing dimensions. The largest, which was installed at the Nazarov opencast (Krasnoyarsk krai) in 1972, is the ESH-80/100, with a boom 100 m. in length and three dippers of 80, 90, and 100 cu.m. capacity. It can remove some 16 million cu.m. of overburden a year. Experts complain that there are not enough large draglines being produced in the USSR, but by 1970 this system was accounting for over 40 percent of opencast production, compared to just 4 percent in 1946.

In the Ukraine a system combining multi-scoop excavators with transport by conveyors and dump-bridges, ("transportno-otvalnaya sistema") is widely used, but in 1970 this system accounted for only 7 percent of opencast production in the USSR as a whole.

TABLE 4.18

Development of Opencast Coal Mining, 1940-72

Year	Opencast Output (in millions of tons)	(As a percentage of total output)
1940	6.3	3.8
1950	27.1	10.4
1960	105.5	20.6
1965	140.5	24.3
1968	150.8	25.4
1969	156.7	25.8
1970	166.6	26.8
1971	171.8	28.0
1972	188.9	29.1

Sources: Melnikov, N. V., Mineralnoe toplivo, Moscow 1971: Ugol no. 4, April 1972: 69; Ugol no. 4, April 1973: 72.

The earliest method, with excavators dumping the overburden into railway wagons or lorries for removal ("transportnaya sistema"), is still in common use, but its share in opencast output has dropped from 95 percent in 1946 to under 50 percent in 1970.

With an overburden of loose soil and rock, a method using water jets and pumps ("gidromekhanizatsiya") has proved successful, but accounts for only about 4 percent of opencast production.[59]

Because it is much cheaper and safer to produce coal by strip mining than from underground, opencast mining is being developed wherever possible in the USSR, even in areas with fairly severe winters.

Opencast reserves at the Moscow Basin have been exploited since the 1950s, and output in future years is unlikely to exceed 3 or 4 million tons annually. The Dnepr Basin is also approaching exhaustion, although the new Verkhnedneprovskoe field, with reserves of 140 million tons, will help to maintain output in this area into the 1980s.

The older opencasts in the Urals are exhausted, and output is steadily dropping. Only in Bashkiria are there sufficient reserves to increase output, and the competition from oil and gas makes this uneconomical at present.

Prospects east of the Urals are much brighter. Kazakhstan has 15,400 million tons of coal suitable for strip mining, concentrated in the Ekibastuz, Turgai, and Maikyubensk basins. Because of the complicated geological structure and large sulphur content of the Turgai coal seams, it is not planned to exploit this basin in the near future, but production of cheap fuel in the other two basins is expanding rapidly. Output in Kazakhstan by strip mining may reach 120 million tons annually by 1985, providing energy for transmission to the Urals and the western USSR.

Central Asia has opencast reserves of only 1,300 million tons, mainly in the Angren area, where output could reach 6 million tons annually.

The Kuznetsk Basin, with reserves of over 9,200 million tons, could produce as much as 100 million tons a year by strip mining.

The Minusinsk Basin, with reserves of 3,500 million tons, could increase output by 1985 to about 8 million tons a year.

The 115,000 million ton reserves of the Kansko-Achinsk Basin guarantee almost limitless quantities of energy coal, up to 1,000 million tons a year at costs of less than a ruble per ton.

The Irkutsk Basin has reserves of 6,800 million tons and could provide 25 million tons a year from opencast mines by 1985.

The Transbaikal fields have reserves of 2,900 million tons, with a potential annual output by opencast methods of up to 20 million tons. The Olon-Shibirskoe field could supply 9 million tons a year of high-calorie hard coal suitable for transporting over long distances.

In the Far East and Sakhalin a large proportion of output at the beginning of the 1970s was still from underground mines, costing three or four times as much as coal from opencast workings. This is expected to change in the near future because there are reserves of over 3,500 million tons suitable for strip mining, and potential output by this method is around 37 million tons a year.

The distribution of reserves of brown and hard coal suitable for opencast exploitation is shown in Table 4.19.

In the USSR an extensive program intended to expand opencast capacity is at present being followed. In 1968 there were 68 opencast mines, with an average capacity of 2,130,000 tons a year; 15 mines were being prepared for exploitation, with an average annual capacity of 5.7 million tons each; and a further 19 were planned that would average 13.5 million tons a year. The simultaneous reconstruction of the older workings has generally doubled their output, and the total result of the program has been to greatly increase average production and to considerably lower the cost per ton. Labor productivity in opencast mining reached 335 tons per man per month in 1972.[60] (See Table 4.20.)

TABLE 4.19

Distribution of Coal Reserves Suitable for
Opencast Exploitation
(in millions of tons)

	Brown	Hard	Suitable for Coking
USSR	139,217	27,205	1,474
European areas and Urals	2,197	—	—
Kazakhstan and central Asia	7,283	7,817	—
Kuzbass	1,224	9,248	1,121
Kansko-Achinsk	119,611	—	—
Minusinsk	—	3,762	—
Eastern Siberia	4,694	5,050	353
Far East and Sakhalin	3,506	73	—

Source: Kuznetsov, K. K., et al., Ugolnye mestorozhdeniya dlya razrabotki otkrytym sposobom, Moscow 1971: 268.

Coal Preparation*

In the USSR considerable efforts are being made to improve the cleaning and dressing of coal. Before 1914 cleaning capacity did not exceed 6 million tons a year. Briquettes were manufactured using imported machinery of a relatively primitive type. The industry did not immediately recover from the civil war, and in 1928 cleaning capacity had not yet reached the prewar level and dressing was still almost completely done by hand.

Several cleaning and dressing works were built in the 1930s, and by 1940 there were 28 factories, mainly in the Donbass, with a yearly capacity of about 26 million tons. The destruction of the factories in European USSR caused by the Second World War was balanced to some extent by the rapid construction of cleaning and dressing works east of the Urals, especially in the Kuzbass.[61]

---

*The Russian term obogashchenie covers the process of separating, concentrating, cleaning, dressing, and washing coal.

Developments in this sector since the war are shown in Table 4.21, and the distribution of factories as of 1965 is shown in Table 4.22.

In recent years some progress has been made in the technology of cleaning and dressing coal. In the 1966-70 period the volume of coal that was prepared by the Coal Ministry of the USSR increased by 50 million tons to 224.7 million tons in 1970. Labor productivity in preparation plants rose by 22.7 percent. During the eighth Five Year Plan, 13 new plants, with a total yearly capacity of 38.4 million tons of "run-of-the-mine" coal, were put into operation. Among the largest were the "Komendantskaya" and "Ayutinskaya" in the Donbass, with 6 million tons yearly capacity; the "Berezovo-Biryulinskaya" in the Kuzbass, with 4.3 million tons; and the "Saburkhanskaya" in Karaganda, with 5.7 million tons. The average yearly capacity of Soviet preparation plants rose from 1.2 million tons in 1965 to 1.5 million tons in 1970.

By 1970 some 56 percent of the preparing of coal had been mechanized. Preparation by flotation had increased from 6.8 percent in 1965 to 8.7 percent, while the use of inefficient rheolaveur washing systems had dropped to 20.6 percent from 31.7 percent in 1965. Although the general quality of "run-of-the-mine" coal has dropped, with the ash content of extracted coal expected to reach 23.2 percent by 1975, the ash content of prepared coal is being held at the 1965 level of about 20 percent. Mechanical methods of separation have been greatly improved.

By 1975 it is planned to increase the preparing of coal to 272.7 million tons in the cleaning plants of the Coal Ministry of the USSR and 344.5 million tons in the country as a whole. Thirteen new plants are being built, and several existing plants are being reconstructed. Average yearly capacity is to reach 1.9 million tons. Labor productivity is expected to increase by 40 percent in the 1971-75 period and reach 650 tons per man per month in 1975.[62]

Transportation

As may be seen in Table 4.23, almost all the transporting of coal is by rail. Every year coal constitutes about a fifth of the total tonnage transported on the Soviet railway system. Because of the great demands made on the railways and the expense involved, only high-quality coal is transported over long distances. The Donbass supplies coal to most of the industrial areas in the center and south of the European USSR, and the Pechora Basin supplies the consumers in the north and northwest. Because it is much cheaper to extract coal at the Kuzbass, its coal and coke are more economical than the coal and coke from the Donbass in many areas of the USSR, in spite

170

TABLE 4.20

Opencast Output and Cost Indices, by Region

	Actual Indices for 1968			Probable Future Trends			
Region	Output (in millions of tons)	Productivity (in hundreds of tons per man per month)	Cost per ton (in rubles)	Output (in millions of tons)	Productivity (in hundreds of tons per man per month)	Cost per ton (in rubles)
European USSR	11.7	2 to 3	2.0 to 4.9	13	3 to 4	2.0 to 3.5
Urals	31.5	1 to 5	1.3 to 5.9	16	2 to 5	1.0 to 2.0
Kazakhstan and Central Asia	24.3	1 to 4	1.5 to 6.7	120	8 to 15	0.5 to 0.9
Kuzbass	25.6	1 to 3	3.5 to 6.5	90	2 to 3	2.2 to 4.0
Krasnoyarsk krai	15.9	2 to 4	1.3 to 5.5	160	7 to 19	0.4 to 0.6
Eastern Siberia	19.9	1 to 7	1.0 to 4.8	40	4 to 7	0.7 to 1.2
Far East	16.0	1 to 5	1.8 to 6.8	40	3 to 7	1.3 to 2.3

Source: Kuznetsov, K. K., et al., Ugolnye mestorozhdeniya dlya razrabotki otkrytym sposobom, Moscow 1971: 273.

TABLE 4.21

Construction of Coal Preparation Plants, 1946-71

	1946	1951	1956	1961	1965	1970	1971
Number of factories	37	68	114	137	155	168	171
Coal processed (in millions of tons)	17.9	50.4	111.6	161.1	214.7	224.7	235.9
Percentage of output	11.4	18.6	26.0	31.7	37.8	36.3	37.4

Sources: Ugolnaya promyshlennost SSSR, 1917-1967, Moscow 1969: 292; Ugol no. 4, April 1971: 73; Ugol no. 4, April 1972: 69.

TABLE 4.22

Regional Distribution of Coal Dressing Works, 1965

Region	Number of Factories	Coal Processed (in millions of tons)	Percentage of total output for area
Donbass and Dnepr	91	128.0	56.5
Kuzbass	33	39.5	53.5
Karaganda Basin	5	8.0	26.0
Urals	6	15.8	31.1
Pechora	10	12.2	67.5
Sakhalin	4	1.8	38.3
Others	6	9.4	29.0

Source: Ugolnaya promyshlennost SSSR, 1917-1967, Moscow 1969: 292.

TABLE 4.23

Transportation of Coal, 1940-71

Mode of Transportation	1940	1950	1960	1965	1970	1971
Rail (in thousands of millions of tons/kilometres)						
Hard Coal	100.8	168.0	318.9	374.4	424.6	441.0
Coke	6.1	10.2	15.0	22.5	23.5	25.9
Rail (in millions of tons transported)						
Hard Coal	145.3	255.0	468.2	552.6	613.9	636.5
Coke	7.3	11.1	24.3	30.4	33.3	33.8
Ship (in millions of tons transported)						
Hard coal (total)	1.8	3.8	7.1	8.2	9.3	8.7
Hard coal, coastal	1.6	2.4	5.7	5.1	4.8	NA*
River, including coke	2.2	4.4	11.0	14.4	17.6	18.6

Sources: Narodnoe khozyaistvo SSSR v 1970 g, Moscow 1971; Narodnoe khozyaistvo SSSR 1922-1972, Moscow 1972.

*NA: not available.

MAP 4.3

Main Coal Flow

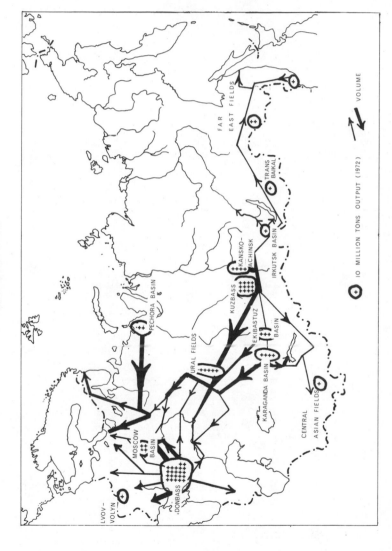

Source: Atlas razvitiya khozyaistva i kultury SSSR, Moscow 1967; Text, Table 4.15.

of the long haulage involved. Coal is shipped from the Kuzbass to the industries of Siberia, eastern Kazakhstan, the middle Urals, and even as far as the central regions of European Russia. The supplying of coal to the steel mills of Magnitogorsk has been of particular importance in the development of the Kuzbass. Karaganda coal is shipped to the southern Urals and to Central Asia.[63]

The construction of high-capacity power stations in Ekibastuz and other bases of cheap coal will allow the more efficient transmission of energy by high-voltage line rather than by rail in the form of coal.

The question of coal transport is closely linked with that of the exploitation of small deposits situated far from railways. It is frequently less expensive to produce even poor-quality coal locally than to transport high-grade coal from large fields over long distances. Even the cheapest Ekibastuz coal, for example, sells for 30 to 50 rubles per ton when it is delivered to the Zaisan area in Eastern Kazakhstan. It would be considerably cheaper to extract coal locally.

The Kenderlykskoe field, 60 km. southeast of Zaisan, had coal and oil shales extracted sporadically from 1894 to 1930 and during the war, but exploitation was not continued because of the distance separating it from the railway, 400 km. away. This very problem of transport, however, has since been seen as a strong argument for reopening this field to provide energy for new local industries.[64]

This policy is also being recommended by planners for some isolated collective farms in Kazakhstan, where two mechanical excavators could be sufficient to produce around 150,000 tons annually for the energy and fertilizer requirements of surrounding farms.

Another transport problem is that of balancing the quality of the coal with the costs of production and delivery. Poor-quality coal produced cheaply by opencast methods can prove to be more expensive as a source of energy than good-quality coal produced by underground mining, if the point of consumption is far removed from the coal field.

An interesting example of this problem is that of supplying the small port of Pevek in the extreme northeast of the USSR with coal from the Zyryanskoe field. While it costs less to produce a ton of coal by opencast methods (6 rubles 83 kopeks) than by underground mining (9 rubles 70 kopeks), the main expense for the Pevek consumer is that of transport (8 rubles 45 kopeks for road transport plus 45 rubles 29 kopeks for sea transport). Because the calorific value of surface coal is only about two-thirds of that mined underground, a greater quantity of opencast coal is required to provide the same energy, and one must therefore compare the costs of the coals in conventional fuel tons after transport charges have been added. This means that at Pevek one ton of conventional fuel from opencast coal would cost 94 rubles 40 kopeks, while one ton of conventional fuel from underground coal would cost much less, 70 rubles 50 kopeks. [65]

The flow of coal from the main fields and basins is shown on Map 4.3. The utilization of coal is discussed in Chapter 7.

NOTES

1. Mining and Technology no. 5, May 1969: 49.
2. "Ugolnaya promyshlennost," Bolshaya Sovetskaya Entsik-
lopedia (Coal industry, Large Soviet Encyclopedia), 2nd edition, Moscow 1947-56; Vol. 43: 626; Ibid., Vol. 50: 172. Ponomarev, B. N.,
et al., Istoriya SSSR (History of the USSR), Moscow 1967, Vol. 4: 29-
33; Bratchenko, B. F., Khleb promyshlennosti (Bread of industry),
Moscow 1967: 5.
3. Ponomarev, op. cit., vol. 7: 305-308, 660-62; ibid., Vol. 8:
517-28; Bratchenko, op. cit.: 6-10; Melnikov, N. V., Mineralnoe
toplivo (Mineral fuel), 2nd ed., Moscow 1971: 100-104; Ugolnaya
promyshlennost SSSR 1917-1967 (Coal industry of the USSR 1917-1967),
Moscow 1969: 4-23; 413-18; Stakhanov, A., "Rodnik rabochikh talantov"
(Spring of working talents), Sotsialisticheskaya industriya, 30 August
1970: 1; Rozhchenko, E. N., "Ugolnaya promyshlennost Rossiiskoi
Federatsii k 50-letiyu obrazovaniya SSSR" (Coal industry of the Russian
Federation on the 50th anniversary of the USSR), Ugol no 12, December
1972: 4-10.
4. Melnikov, op. cit.: 23; Energeticheskie resursy SSSR:
toplivno-energeticheskie resursy (Energy resources of the USSR:
fuel resources), Moscow 1968: 36.
5. Ibid.: 62; Chemical and Engineering News, 13 November
1972: 21.
6. Energeticheskie resursy, op. cit.: 63-64.
7. Ibid.: 67.
8. Kravtsov, A. I., Trofimov, A. A., "Itogi Vsesoyuznogo
nauchno-tekhnicheskogo soveshchaniya geologov ugolnoi promyshlennosti" (Achievements of the All-Union scientific-technical conference
of geologists of the coal industry), Ugol no. 3, March 1972: 68-69.
9. Energeticheskie resursy, op. cit.: 68-78; Ugol no. 4, April
1973: 72.
10. Energeticheskie resursy, op. cit.: 78-83; Ugol no. 4, April
1973: 72.
11. Energeticheskie resursy, op. cit.: 78-83; Ugol no. 4, April
1973: 72; Kuznetsov, K. K., et al., Ugolnye mestorozhdeniya dlya
razrabotki otkrytym sposobom (Coal fields for opencast exploitation),
Moscow 1971: 17-21; Ekonomicheskaya gazeta no. 8, February 1973:
10.
12. Energeticheskie resursy, op. cit.: 85.
13. Ibid.: 86; Kuznetsov, op. cit.: 21-30.
14. Ibid.: 30-32; Energeticheskie resursy, op. cit.: 86-87.
15. Ibid.: 87-89.
16. Ibid.: 89-90.
17. Ibid.: 80-91.

18. Ibid.: 91-92; Kuznetsov, op. cit.: 43-46.
19. Ibid.: 46-49; Energeticheskie resursy, op. cit.: 92-93.
20. Ibid.: 93; Kuznetsov, op. cit.: 33-43.
21. Energeticheskie resursy, op. cit.: 93.
22. Ibid.: 49-51; Tarkhaneev, B., "S positsii gosudarstvennykh, a ne vedomstvennykh" (Putting the State before the department), Sotsialisticheskaya industriya, 3 September 1970: 2; Melnikov, N., et al., "Ischerpany li klady?" (Is the treasure exhausted?), Pravda, 16 August 1972: 3.
23. Energeticheskie resursy, op. cit.: 95; Kushev, G. L., Ugolnye bogatstva Kazakhstana (Coal riches of Kazakhstan), Alma Ata 1971: 5; Ugol no. 12, December 1972: 21.
24. Energeticheskie resursy, op. cit.: 95-99; Ugol no. 4, April 1973: 72.
25. Energeticheskie resursy, op. cit.: 99; Kuznetsov, op. cit.: 65-66.
26. Ibid.: 56-61; Kushev, op. cit.: 25-29; Sotsialisticheskaya industriya, 13 January 1973: 2; Sotsialisticheskaya industriya, 16 January 1974: 2; Energeticheskie resursy, op. cit.: 99-100.
27. Ibid.: 101; Kuznetsov, op. cit.: 61-64.
28. Ibid.: 66-74; Shortanbaev, A., "Gde vzyat ugol dlya Urala?" (Where to get coal for the Urals?), Sotsialisticheskaya industriya, 3 September 1970: 2; Kushev, op. cit.: 32; Energeticheskie resursy, op. cit.: 101-102.
29. Ibid.: 102-103.
30. Ibid.: 103-105. Taskaev, V. V., "Ugolnaya promyshlennost Srednei Azii v devyatoi pyatiletke" (The coal industry of central Asia in the ninth Five Year Plan), Ugol no. 12, December 1972: 21-24; Kuznetsov, op. cit.: 82-95.
31. Ibid.: 96-147; Energeticheskie resursy, op. cit.: 105-11; Ugol no. 4, April 1973: 72; Pravda, 12 April 1973: 2.
32. Energeticheskie resursy, op. cit.: 111.
33. Ibid.: 112-15; Kuznetsov, op. cit., 148-66; Melnikov, op. cit.: 23; Semikobyla, G. S., "Aktivnee razvivat dobychu uglya v Krasnoyarskom krae" (Expanding the extraction of coal in Kraznoyarsk territory more energetically), Ugol no. 3, March 1972: 4-12.
34. Ibid.: 4-12; Energeticheskie resursy, op. cit.: 115-16; Kuznetsov, op. cit.: 166-75.
35. Ibid.: 185-90; Energeticheskie resursy, op. cit.: 116.
36. Ibid.: 116-17; Kurenchanin, V. K., Razrabotka ugolnykh mestorozhdenii severo-vostoka SSSR (Exploitation of coal fields in the northeast USSR), Moscow 1971: 126, 132; Kuznetsov, op. cit.: 176-85.
37. Ibid.: 191-208; Energeticheskie resursy, op. cit.: 117-20.
38. Ibid.: 12-123; Kuznetsov, op. cit.: 209-228.

39. Ibid.: 250-53; Energeticheskie resursy, op. cit.: 123-24.

40. Kurenchanin, op. cit.: 19-20: 35.

41. Ibid.: 79-86; Kuznetsov, op. cit.: 256-59.

42. Ibid.: 259-62; Kurenchanin, op. cit.: 111; Energeticheskie resursy, op. cit.: 129.

43. Kuznetsov, op. cit.: 262-64.

44. Ibid.: 246; Kurenchanin, op. cit.: 67-78; Energeticheskie resursy, op. cit.: 124-25.

45. Ibid.; 125-26; Kuznetsov, op. cit.: 234-39.

46. Energeticheskie resursy, op. cit.: 126-27.

47. Ibid.: 128; Kuznetsov, op. cit.: 232-33.

48. Energeticheskie resursy, op. cit.: 127.

49. Ibid.: 127-28.

50. Ibid.: 128-29; Kuznetsov, op. cit.: 230-34.

51. Ibid.: 244-45; Energeticheskie resursy, op. cit.: 129-30.

52. Melnikov, op. cit.: 138; Ugol no. 3, March 1972: 72.

53. Narodnoe khozyaistvo SSSR v 1970 godu (Economy of the USSR in 1970), Moscow 1971: 189; Melnikov, op. cit.: 105-108.

54. Ibid.: 117-18; Ugol no. 4, April 1972: 71; Energeticheskie resursy, op. cit.: 134-67; Ugol no. 12, December 1972: 9; Trud, 13 January 1973: 2.

55. Sotsialisticheskaya industriya, 30 August 1970: 2.

56. Ekonomicheskaya gazeta no. 37, September 1972: 4.

57. Ugolnaya promyshlennost SSSR, op. cit.: 231-234.

58. Serechenko, A. A., et al., "Mekhanizatsiya vspomogatelnykh protsessov na shakhtakh" (The mechanization of auxiliary processes in mines), Ugol no. 5, May 1972: 62-64; "Novyi vazhnyi etap v razvitii ugolnoi promyschlennosti SSSR" (New important stage in the development of the USSR coal industry), Ugol no. 1, January 1974: 4-7; Trud, 3 January 1974: 2; Trud, 17 April 1973: 1.

59. Melnikov, op. cit.: 118-29; Energeticheskie resursy, op. cit.: 167-80; Pravda, 10 August 1972: 2.

60. Kuznetsov, op. cit.: 267-74.

61. Ugolnaya promyshlennost SSSR, op. cit.: 288-93.

62. Blagov, I.S., "Zadachi ugleobogatitelei v devyatoi pyatiletke" (Coal-dressing tasks in the ninth Five Year Plan), Ugol no. 1, January 1972: 60-64.

63. Khrushchev, A. T., Geografiya promyshlennosti SSSR (Industrial Geography of the USSR), Moscow 1969: 186-89.

64. Kushev, op. cit.: 36-45.

65. Kurenchanin, op. cit.: 85-86.

5

Although by the beginning of the 1970s peat had dropped to under 2 percent of the total Soviet fuel balance, the peat industry still retains two very important functions.

In parts of the USSR deficient in high-quality fuels (oil, gas, coal), it is often more economical to obtain energy from power stations fired by local peat than to use expensive fuels transported over long distances. In 1960, for example, peat supplied half of the energy requirements of the Belorussian SSR and over two-thirds those of the Ivanovo and Kalinin oblasts.

Agriculture is the main consumer of the peat produced in the USSR. The demand for fertilizers has been growing steadily and is not yet being fully satisfied. Bedding for cattle, litter for poultry, and packing for fruit and vegetables also use large quantities of peat.

The Soviet peat industry is therefore expected to continue to expand, although the share of peat in the total fuel balance will eventually drop to less than 1 percent.

HISTORICAL BACKGROUND

In Russia the first attempts to organize the production of peat on a large scale were made in the 18th century. A Senate decree on this subject was signed by Peter the Great in 1723, and with some government encouragement output increased slowly through the 19th century. In 1851, for example, a government committee "for expanding the utilization and exploitation of peat" was formed in Moscow. Intensive production took place at the Paltso bog near Bryansk and in the areas between the rivers Oka and Klyazma. Peat was an important industrial fuel in the textile factories around Moscow. By 1900 about 1.6 million tons were being extracted annually in Russia.

In 1914, when 1.7 million tons were produced, the first peat-fired power station was commissioned. Situated near Moscow, it had a maximum capacity of 15,000 kW.

Output fell to 1.4 million tons in 1916, but during the Civil War, when the Center was cut off from the oil of the Caucasus and from Donbass coal, particular attention was paid to developing local resources of peat. In 1918 exploitation of the Shatura fields near Moscow began. The State Commission for the Electrification of Russia (GOELRO) plan (1920) called for peat output by 1930 to be ten times what it had been in 1916. Out of thirty thermal power stations planned, four were to be peat-fired with an aggregate capacity of 170,000 kW. These were built in Nizhni Novgorod (Gorkii), Ivanovo-Voznesensk, Petrograd (Leningrad), and Shatura.

Former methods of peat extraction by backbreaking work using spades were no longer efficient enough. To raise production and avoid seasonal restrictions new mechanical methods had to be developed. In 1919, P. E. Klasson invented a hydraulic method of excavating and transporting peat that by 1924 was responsible for 188,600 tons out of the USSR total of 2,857,000 tons. In 1940 as much as 37 percent (9,300,000 tons) of the peat produced in the RSFSR was by this hydraulic method. This form of peat winning had gone out of use by 1963, having produced a total of 187 million tons. In the same period other machines were developed for the excavation and elevation of peat from the bog.

In 1928 a new process was applied that fully mechanized all the operations: cutting, drying, collecting, transporting, and stacking. This process, called milling, known in Russian as the "frezernyi sposob", entails the use of milling machines that cut the peat into easily managed small pieces. The surface of the bog is cut to a depth of 1 to 2 cm. over its entire area by rotating drums on which small cutting pins are fixed. These drums were originally drawn by tractors, but there are now several different types of self-propelled milling machines in operation. The powdered peat is left to dry on the surface and later harrowed to speed up drying; its moisture content drops from 90 percent to below 50 percent. The dried peat is gathered into rows, lifted by harvesting machines, and formed into piles alongside a narrow-gauge railway for collection.

In 1932 some 13.5 million tons of peat were produced, and in 1937 output reached 24 million tons. Mechanization of the peat industry rose to almost 50 percent by 1940, when 33.2 million tons were extracted. As much as 20 percent of the electricity produced in that year was from peat-fired stations.

The Second World War emphasized the importance of the peat industry as a reliable source of local fuel. In the Leningrad, Kalinin, Gorkii, and Sverdlov oblasts peat was actually the main source of energy. Although output fell sharply because of the occupation of much of the main productive territory in the west, recovery was rapid; and by 1950 the prewar level of extraction had been surpassed.

Mechanical and hydraulic methods of extraction were improved, but the greatest progress was made by the complex milling process. While in 1940 milling had been responsible for 16 percent of production, by 1958 it accounted for 53 percent, rising to 93 percent in 1970. The peat industry was now less dependent on good summer weather and could continue operations for a greater part of the year.

However, the climate in peat-winning areas is still a decisive factor in determining the success of operations. In the decade from 1956 to 1965 inclement weather frequently caused considerable loss of production, especially in the new northern fields. Output in 1965, for example, was some 13.5 million tons below the 1964 level. The planned figure for the production of peat fuel in 1970 was originally 75.6 million tons, but this was lowered to 60 million tons, and in actual fact only 57.3 million tons were produced, officially because of unfavorable weather in Belorussia and the western Ukraine. In August 1972 widespread peat and forest fires disrupted production in the central area near Moscow. (There was scarcely any official mention of this until an indirect reference to the extent of the damage appeared in Pravda on 19 October 1972.)

Nonetheless, the growth in output in the Soviet period has been remarkable; the average yearly increase from 1917 to 1927 was 350,000 tons; from 1928 to 1940 the yearly increase rose to 2.1 million, and immediately after the drop in production caused by the war it reached a maximum growth rate of 3.8 million tons for 1946-48. Since then, in spite of considerable fluctuations, the average annual growth has been about 2 million tons, more than the total yearly output in tsarist times. In 1972 peat enterprises produced 80.6 million tons for fuel and agriculture, some fifty times the production in 1913.[1]

It is much more difficult to provide accurate and comprehensive statistics on the development of the peat industry than it is for other fuels. Soviet sources do not always state the water content when giving production figures in tons, and while one can therefore expect some discrepancies in different tables, it is difficult to explain why the figures should coincide for some years, yet be totally different for others. Thus, sources A, B, and C in Table 5.1 agree that in

---

*Academician Melnikov would appear to be mistaken in including 72.2 million tons as fuel peat output in 1965. A later table in his own book (p. 165) shows this figure for the total output including peat for agriculture, while giving 55.4 million tons for fuel peat. Yet even this amount is greater than the figure for 1965 in other sources.

TABLE 5.1

Fuel Peat Production in USSR, 1913-71

Year	Output[a] (in millions of tons)	Output[b] (in millions of tons of conventional fuel)	Percentage of Fuel Balance[b]
1913	1.7	0.7	1.4
1917	1.4	NA	NA
1918	1.1	NA	NA
1928	5.3	NA	4.0
1932	13.5	NA	5.2
1937	24.0	NA	5.3
1940	33.2	13.6	5.7
1941	27.0	NA	NA
1945	22.4	9.2	4.9
1946	27.2	11.2	5.5
1950	36.0	14.8	4.8
1955	50.8	20.8	4.3
1957	53.7	22.5	3.9
1958	52.5	21.1	3.4
1959	59.4 (60.5[c])	23.0	3.5
1960	52.9 (53.6[d])	20.4	2.9
1961	51.0	19.5	2.7
1962	34.7	12.9	1.7
1963	57.9	21.7	2.5
1964	58.3	22.2	2.4
1965	45.1 (55.4[e])	17.0	1.7
1966	66.4	24.4	2.3
1967	NA	22.4	2.1
1968	NA	18.3	1.6
1969	44.8	16.7	1.4
1970	57.3[d]	21.4	1.7
1971	54.8[f]	16.7[g]	1.3[g]

NA: Not available.

[a]Most likely figures (Source A), with alternative figures in parentheses.
[b]Source B. [c]Source C.
[d]Source D. [e]Page 165 of Source E.
[f]Source F. [g]Source G.

Sources: Source A: Torf v narodnom khozyaistvo, Moscow 1968: 59; Source B: Narodnoe khozyaistvo SSSR v 1970 g. Moscow 1971: 183; Source C: Narodnoe khozyaistvo SSSR v 1960 g. Moscow 1961; Source D: Torfyanaya promyshlennost no. 2, 1971: 6; Source E: Melnikov, N. V., Mineralnoe toplivo Moscow 1971: 148, 165; Source F: SSSR v tsifrakh v 1971 g. Moscow 1972: 86-87; Source G: Narodnoe khozyaistvo SSSR 1922-1972 Moscow 1972.

1928 about 5.3 million tons were produced for fuel, and in subsequent years the same sources generally show variations of under 1 million tons; yet for 1965 A states that 45.1 million tons were produced for fuel, B's figure is 45.7 million tons, and C claims 72.2 million tons! One might consider the only completely reliable figures to be those of the official statistical yearbook, Narodnoe khozyaistvo SSSR, which shows peat output in tons of conventional fuel. Even here, however, it must be noted that the fluctuations in output figures do not invariably match the pattern elsewhere. While other sources show that about 2 million tons more peat were extracted in 1960 than in 1955, we find that in conventional fuel tons production was actually lower in 1960 than in 1955. The conventional fuel figure for 1971 would imply that production was actually 10 million tons less than the 54 million tons claimed. There is probably a logical explanation for these discrepancies, but confusion could be avoided by more explicit information made available in tables and figures. Table 5.1 shows the most likely output for selected years since 1913, with alternative figures in parentheses. The latter possibly include fuel peat extracted by state farms (sovkhozes) in addition to that extracted by the peat and fuel ministries. It should be borne in mind that more peat is extracted by agencies involved in agriculture than by the peat enterprises discussed in this chapter. Statistics refer only to output by these enterprises.

PRESENT AND FUTURE DEVELOPMENT

Location of Reserves

World reserves of peat are estimated at some 261,400 million tons (with 40 percent water content), of which the USSR has about 158,000 million tons, or 60 percent of the total. Other countries with large peat reserves are Finland, with 12 percent of the world total; Canada, with 9 percent; and the United States, excluding Alaska, with 5 percent.[2]

Soviet peat reserves are situated mainly in the Northwest, the northern Urals, western Siberia, Kamchatka, and Sakhalin. As much as 92.1 percent of the total geological reserves of 158,000 million tons is situated in the RSFSR, 3.4 percent in Belorussia, and the remainder in the Ukraine and Baltic republics. Within the RSFSR, peat reserves are located mainly in the north (10.4 percent), northwest (5.6 percent), center (4.2 percent), the Urals (5.2 percent), and Western Siberia (56.6 percent).[3]

Legend

A   Belorussia

B   Latvia

C   Estonia

D   Leningrad area

E   Moscow area

F   Upper Volga area

G   North West

H   Komi ASSR

I   West Urals

J   Middle Urals

K   Omsk area

L   Far East

MAP 5.1

Peat Reserves

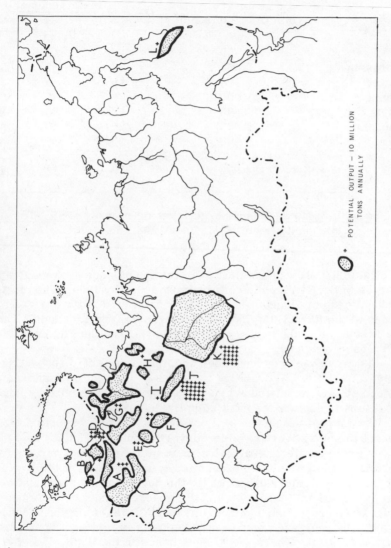

POTENTIAL OUTPUT — 10 MILLION
TONS ANNUALLY

Sources: Torfyanaya promyshlennost no. 4, April 1970; Energeticheskie resursy SSSR.— Toplivo-
energeticheskie resursy, Moscow 1968: 552; Text, Table 5.3.

TABLE 5.2

Distribution of Peat Reserves and Extent of
Exploration, by Republic

Republic	Geological Balance Reserves (in millions of tons)	Intensity of Exploration (in percentage)		
		A + B	C_1	C_2
RSFSR	111,700	6	10	84
Belorussia	5,400	28	35	37
Ukraine	2,300	18	30	52
Estonia	1,900	6	37	57
Latvia	1,600	19	56	25
Lithuania	1,000	20	40	40
Others	100	50	50	—
USSR	124,000	8	13	79

Source: Adapted from Energeticheskie resursy SSSR: Top-livno-energeticheskie resursy, Moscow 1968: 552.

Table 5.2 shows the extent to which the balance geological peat reserves of Soviet republics had been explored by 1966. The most thoroughly explored and exploited areas are the Moscow and Volga regions of the RSFSR, the Ukraine, Belorussia, and Lithuania. The huge resources of the Urals, western Siberia, and the Far East remain almost untapped.

Many new methods have been adopted to simplify prospecting for peat resources. Calculating the sizes of the peat bogs and mapping the deposits are mainly done by aerogeology, supplemented by land expeditions to explore the most complicated areas.

Not all peat resources, and not even all of the proportion termed "balance reserves," are considered worth exploiting unless they satisfy various prerequisites. Peat fields must be large enough in area for industrial exploitation to be economically viable; they must be situated near industrial centers or railways; the quality of the peat must conform to the official standards; and the reserves must be sufficient to guarantee production continuing at least 30 years. If these conditions are met, the approved group of peat deposits is termed a "peat base" (torfyanaya baza). There are 77 such bases in the USSR, 65 in the RSFSR, 8 in Belorussia, and 1 each in the Ukraine, Lithuania, Latvia,

TABLE 5.3

Location of Peat Reserves and Potential Production

Area	Number of Peat Bases	Exploitable Reserves (in millions of tons of air-dried peat, water content of 40 percent)	Potential Annual Output (in millions of tons)
Belorussia	8	1,007.4	33.3
Ukraine	1	67.0	2.2
Lithuania (with Kaliningrad)	2	330.0	2.3
Latvia	1	900.0	7.0
Estonia	1	210.0	7.0
RSFSR			
Leningrad	11	2,385.4	79.2
Moscow	6	872.7	29.1
Priokskii	2	59.0	1.9
Upper Volga	8	629.0	20.7
Volga-Vyatskii	6	406.9	13.6
Northwest	10	719.7	23.2
Komi ASSR	4	410.6	13.7
West Urals	2	402.0	13.4
Middle Urals	9	8,705.2	288.5
Western Siberia	3	10,740.0	214.8
Far East	3	448.2	14.9
USSR	77	28,293.1	764.8

Sources: Energeticheskie resursy SSSR: Toplivno-energeti-cheskie resursy, Moscow 1968: 555-61; Torf v narodnom khozyaistve, Moscow 1968: 58.

and Estonia.[4] The location of peat reserves and of potential production is shown in Table 5.3 and on Map 5.1.

It is possible to divide the Soviet Union into four zones when discussing reserves and production. (See Map 5.2.) Zone 1 includes the Moscow, Ivanovo, Bryansk, Smolensk, and Ryazan oblasts. Many important peat bases have been developed here because of their favorable location near industrial centers or rail and river communication

MAP 5.2

Main Peat Producing Areas

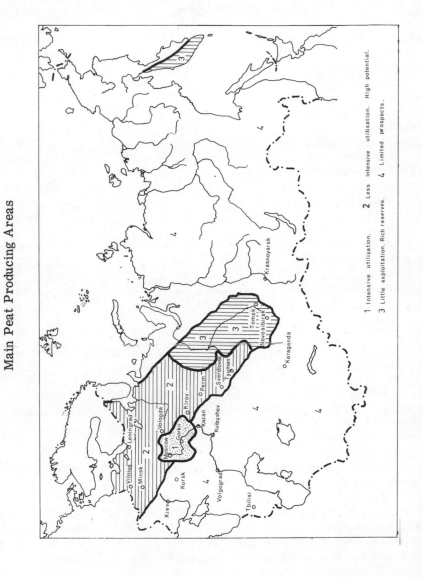

1 Intensive utilisation.   2 Less intensive utilisation. High potential.

3 Little exploitation. Rich reserves.   4 Limited prospects.

Source: Matveev, A. M. et al., Torf v narodnom khozyaistve, Moscow 1968.

systems and also because of their ease of exploitation and relatively high quality as a fuel. Over 60 percent of the resources of this zone have been brought into production. Out of balance reserves of about 2,000 million tons, some 30 million tons are extracted every year; this accounts for about a fifth of the total annual production of the USSR, including peat for agriculture. For every 100 million tons of reserves, about 1.5 million tons are extracted yearly, compared to the Soviet average of only .13 million tons.

Zone 2 is the main area for the further expansion of the peat industry until 1990. It covers the rich peat-bearing territory to the west, north, and east of the first region and includes the Urals and the western and northwestern regions. Although explored reserves are conveniently situated and over twenty times in volume those of Zone 1 (about 48,000 million tons), annual output by the beginning of the 1970s was only 110 to 120 million tons; this could be increased to over 350 million tons a year. Among the most promising peat bases in the northwestern part of the USSR are the Cherepovetskaya, Polistov-skaya, Tikhvinskaya, Orshinskaya, and Zharkovsko-Svitskaya; here the peat is compact, deep, and of good quality. The peat industry is already well established here, but it is not developed sufficiently for local energy requirements. Several large-capacity power stations are being built, and utilization of peat for briquettes, chemicals, cattle bedding, fertilizers, and insulation will be expanded. Among the promising peat bases of the Urals are the Tavdinskaya, Taboro-Lenskaya, Verkhoturskoe-Sinichikhinskaya, and Ivdel-Garinskaya. In the western part of Zone 2, including Belorussia, the southern territory of the Baltic states, and the northwestern Ukraine, there are high quality explored reserves of 8,000 million tons, which will be utilized mainly for energy purposes.

Zone 3 includes two of the richest peat-bearing areas of the USSR, the West Siberian lowlands (without the southwestern part of Tyumen oblast) and Kamchatka. There has been little exploitation because of unfavorable economic and natural conditions, since most of the peat is located in inaccessible swamps far from centers of population. Some peat is consumed by the Tyumen power station, and in Kamchatka peat is extracted in small quantities for fertilizers. In 1971, after research had shown that by replacing only half of the imported coal by local peat it would be possible to save 11 million rubles a year, preparatory work for large-scale extraction of Kam-chatka peat began on the western coast. Annual output may be brought to 20 million tons by 1975.

Zone 4 covers the less favorable areas of the southern Ukraine and RSFSR, Transcaucasia, the Far East, and the Far North, which have nothing in common climatically or in types of peat, but do share an absence of concentrated reserves that could serve as the basis for large-scale peat industries. In Georgia peat has been extracted in small quantities for fertilizer for cultivating citrus fruits and tea.[5]

It is worth examining the industrial potential of some examples of peat bases in detail. Among the most important plans for the immediate future are the utilization of several peat bases for the building of power stations.

The Polistovskaya base in Novgorod oblast is made up of 13 peat bogs, with reserves of 335 million tons, capable of yielding 11.2 million tons annually. A power station of 600 MW capacity is planned here.

The Cherepovetskaya base in Vologda oblast joins 37 peat bogs with reserves of 668 million tons and a potential annual yield of 22.3 million tons. This could include 12 to 15 million tons of peat fuel, 4 to 5 million tons of peat fertilizer, 1 million tons of cattle bedding, and 7 million sq.m. of insulation tiles. Moreover, since most of the reserves of peat have no sulphur content and little ash (2.7 percent), it would even be possible to use peat in the metals industry. Reserves suitable for coking could be used in the production of up to 4 million tons of cast iron a year. A power station of 600 MW capacity is already in operation.

The Tavdinskaya base in Sverdlovsk oblast includes 18 peat bogs, with reserves of 883 million tons and a potential yearly output of 17.8 million tons. This could support power stations with an aggregate capacity of 3,000 MW. The peat is of high quality and could be utilized for metals processing, cattle bedding, insulation, concentrated fertilizers, and chemical processing. According to some calculations, yearly production could rise to 80 million sq.m. of insulation tiles, 6 million tons of cattle bedding, 1.5 million tons of coke, 25 million tons of energy and industrial fuel, and 20 million tons of organic fertilizer.

In Western Siberia there are some 90,000 million tons of geological peat reserves, 60 percent of the total for the USSR. This could provide sufficient fuel for power stations with an aggregate capacity of over 20,000 MW.[6]

With explored reserves of over 120,000 million tons, the Soviet Union could increase the total annual production of peat for all purposes up to a level of 1,000 million tons. Although there is at present no necessity for such a huge output, research into the properties and possible uses of peat is continuing, and the consumption of peat for agriculture or to produce energy is growing steadily.[7]

Production

The USSR is the major peat-producing country in the world; in 1967, for example, over 22 million conventional fuel tons of peat were produced in the USSR, while the world total was about 26 million tons. Because of the great expense involved in transporting coal, oil, or natural gas to many remote areas, locally extracted peat often proves to be the most economical fuel. The republics with the highest output are the RSFSR, the Ukraine, Belorussia, and the Baltic States, as may be seen in Table 5.4 and Figure 5.1. Some 200,000 to 300,000 tons is produced each year in Georgia and about 10,000 tons in Armenia, but this is used only for agricultural purposes. In the 1940s and early 1950s small quantities were also produced in the Central Asian republics.[8]

In addition to being extracted for fuel, considerable quantities of peat are used in agriculture. In 1964, for example, total output reached 292.7 million tons, of which individual sovkhozes extracted 70.2 million, local soviets 2.5 million, and agricultural ministries and departments 169.7 million, with the remainder being produced by the peat enterprises proper.[9]

Peat has a competitive position in the fuel balance of many areas in the USSR; this is mainly possible because the extraction process is almost completely mechanized. A wide variety of machines have been developed to excavate drainage ditches and to cut, stack, and transport peat. Over 93 percent of the output of peat enterprises is

TABLE 5.4

Output of Fuel Peat, by Republic, 1928-71
(in thousands of tons)

Year	USSR (total)	RSFSR	Ukraine	Belo-russia	Lithuania	Latvia	Estonia
1928	5,320	5,087	144	84	—	—	—
1932	13,495	11,760	902	834	—	—	—
1937	24,040	19,949	1,552	2,529	—	—	—
1940	33,229	25,569	3,544	3,361	102	213	283
1950	35,999	27,490	2,928	3,912	505	623	476
1955	50,777	36.069	4,119	7,191	1,595	1,266	502
1960	53,625	36,816	4,664	8,312	1,554	1,795	467
1965	45,747	29,402	4,343	8,366	1,248	1,649	670
1970	57,300	39,449	4,030	9,232	1,481	2,148	974
1971	54,300	33,700	4,523	11,252	1,492	2,285	1,048

Sources: Torfyanaya promyshlennost no. 2, 1971: 6-7; Narodnoe khozyaistvo SSSR 1922-1972, Moscow 1972.

FIGURE 5.1

Output of Fuel Peat by Republic, 1950-1971

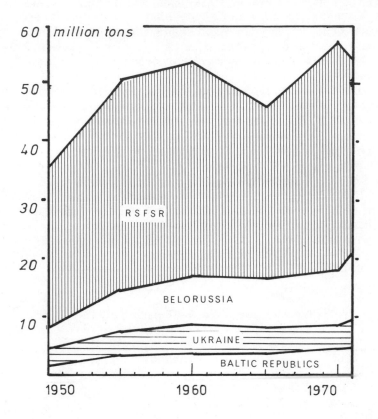

Source: Table 5.4.

by milling machines. Several types of self-propelling pneumatic combines are now in use.10

Peat is normally utilized in close proximity to the point of extraction, and transport is mainly by light narrow-gauge railway (750 mm.). This is often of a temporary nature, constructed by special track-laying trains that use a minimum of ballast. Several thousand km. are laid every year; the longest stretches are about 120 km., and the average is about 20 km.11

The increased use of milling methods of extraction has simplified the production of peat briquettes. Milled peat is passed through drying pipes at temperatures up to 700°C and then compressed. Water content is reduced to 12 to 15 percent, compared to a peat fuel norm of 40 to 45 percent. Briquettes have a calorific value of about 4,500 k.cal./kg. Expansion in the output of peat briquettes in the USSR since 1950 is shown in Table 5.5. An increase from 5.3 to 15 million tons in the present decade seems somewhat optimistic, but considerable growth can nonetheless be expected.12

Utilization

Peat extraction enterprises account for about a third of the total output in the USSR. (The remaining two-thirds is used directly for

TABLE 5.5

Production of Peat Briquettes in USSR, 1950-80
(in thousands of tons)

Area	1950	1960	1970	1980 (planned)
USSR	350	1,458	5,300	15,000
RSFSR	269	799	1,750	7,900
Ukraine	2	133	770	1,200
Belorussia	17	403	2,000	5,000
Lithuania	0	13	320	600
Latvia	7	10	170	600
Estonia	55	100	290	500

Sources: Torfyanaya promyshlennost no. 2, 1971: 7; Melnikov, N.V., Mineralnoe toplivo, Moscow 1971: 165.

agricultural purposes and is not dealt with here.) Available statistics for the 1965-70 period break down the consumption of peat by various sectors of the economy as follows: electric power stations 52 percent; industrial boilers 5 percent; communal domestic heating 8 percent; briquette production 15 percent; agriculture 20 percent. The peat-fired power stations are mainly in the Northwest and the Urals. In 1972 the second unit of the Shatura power station (750 MW) was completed, and work continued on the Cherepovetsk, Pskov, and Smolensk power stations, which have a planned capacity of 600 MW each. The production of briquettes is economically viable only in the European part of the USSR, in areas where the demand for fuel is sufficient to justify transport over short distances.

In the future peat could be improved as an energy fuel by enriching it with heavy fuel oil. Peat may also be increasingly used in metals processing because of its low sulphur and phosphorus content, especially since it is often to be found close to iron ore deposits and metals plants.[13]

It is a useful raw material for the chemical industry. Processing yields commercial quantities of ethyl alcohol, oxalic acid, furfurol, alimentary yeasts, antiseptics, plastics, dyes, varnishes, glues, and many other useful substances.[14]

In 1971 some 200,000 cu.m. of thermal insulation tiles were produced from peat, and this use is expected to grow steadily in future years.[15]

Although it seems likely that the annual output of peat in the USSR will continue to fluctuate widely, reserves are sufficient to allow considerable increase in production. The main consumer will still be agriculture, but the construction of peat-fired power stations and associated chemical plants is being planned in areas with large peat reserves.

While remaining a comparatively insignificant part of the Soviet economy as a whole, the peat industry will therefore continue to be of local importance in many areas of the USSR.

NOTES

1. Matveev, A. M., et al., Torf v narodnom khozyaistve (Peat in the national economy), Moscow 1968: 7-10; Melnikov, N. V., Mineralnoe toplivo (Mineral fuel), Moscow 1971: 147; Pervukhin, A. G., "Pyatidesyatiletie Gosplana SSSR i razvitie torfyanoi toplivnoi promyshlennosti" (Fiftieth anniversary of Gosplan and the development of the peat fuel industry), Torfyanaya promyshlennost no. 2, February 1971: 2-6; "Torfyanaya promyshlennost" (Peat industry), Bolshaya Sovetskaya Entsiklopediya, 2d. ed., Moscow 1947-56: vol 43;

Ivashechkin, N. V., "Torfyanaya promyshlennost RSFSR" (Peat industry of RSFSR), Torfyanaya promyshlennost no. 12, December 1972: 2-8.

2. Olenin, A. S., "Torfyanye resursy SSSR," Torf v narodnom khozyaistve (Peat resources of the USSR, Peat in the national economy), Moscow 1968: 43; Melnikov, op. cit.: 147.

3. Energeticheskiye resursy SSSR: Toplivno-energeticheskie resursy (Energy resources of the USSR: fuel resources), Moscow 1968: 549; Melnikov, op. cit.: 148.

4. Olenin, op. cit.: 48; Energeticheskiye resursy, op. cit.: 552.

5. Olenin, op. cit.: 50-57; Torfyanaya promyshlennost no. 1, January 1971: 15-16; Torfyanaya promyshlennost no. 6, June 1972: 4-5.

6. Energeticheskiye resursy, op. cit.: 560-62; Matveyev, A. M., et al., "Perspektivy razvitiya torfyanoi promyshlennosti" (Prospects for the development of the peat industry), Torf v narodnom khozyaistve, Moscow 1968: 250-52.

7. Energeticheskiye resursy, op. cit.: 563.

8. Lushnikov, A. A., "Ispolzovanie torfa dlya toplivno-energeticheskikh tselei" (Utilization of peat for energy purposes), Torf v narodnom khozyaistve, Moscow 1968: 60.

9. Melnikov, op. cit.: 148.

10. Ibid. 155-63; Matveev, op. cit.; Sotsialisticheskaya industriya, 11 March 1971: 2.

11. Grachev, V. A., Fedorov, V. V., "Transport torfa" (Peat transport), Torf v narodnom khozyaistve, Moscow 1968: 202-203.

12. Melnikov, op. cit.: 164.

13. Matveev, op. cit.: 253-55; Melnikov, op. cit.: 150.

14. Sopin, P. F., "Proizvodstvo khimicheskikh produktov iz uglevodov torfa" (Chemical products from peat carbohydrates), Torf v. narodnom khozyaistve, Moscow 1968: 183-202.

15. Torfyanaya promyshlennost no. 6, June 1972: 2-3.

In August and September 1968 a United Nations symposium on the development and utilization of oil shale* resources was held in Tallin, the capital of the Estonian SSR. In their final recommendations the participants stated that they were convinced "that oil shale resources are an important potential raw material, which with further development could offer a significant contribution to the economy of many countries."[1]

While oil shale can be commercially distilled to produce liquid fuels, in the USSR it is at present more important as a local source of fuel that can be utilized without costly processing in thermal power stations. It is even less important in the total Soviet energy balance than peat, since it provides less than one percent of the nation's fuel requirements.

HISTORICAL BACKGROUND

Oil shale was being extracted on a small scale over 600 years ago in Manchuria, but its commercial distillation was first practised in France in 1838. Scotland began to utilize oil shale in significant quantities in 1859, with output reaching a peak of almost 4 million tons during the First World War.

It was also at this time that shale first took on importance in Russia, where previously it had only been extracted by primitive

---

*The commonly accepted English term "oil shale" has been retained as an equivalent of the Russian "goryuchii slanets" (combustible shale).

methods for domestic use. The fuel crisis brought on by war encouraged the utilization of Estonian shale to supply energy to the St. Petersburg industrial region. In 1916 some 1,250 tons of shale were mined in Estonia and sent to St. Petersburg. While shale was far below hard coal in quality, it was found possible to use shale in solid form in factory power generators and even in railway locomotives, although the boilers had to be cleaned more frequently.

In July 1918 the new Soviet government passed a decree calling for the immediate exploration and development of oil shale resources in the Baltic area, which was under its control. During the Civil War the region with the greatest known resources of oil shale, the Baltic Basin, was the scene of bitter fighting between pro-Soviet and Estonian nationalist forces; when hostilities ceased, Estonia, Latvia and Lithuania were recognized as independent republics.

The shale industry expanded considerably in Estonia between the wars. The first large consumers of oil shale were cement factories; in 1922 the Port-Kunda and Aseri factories in Estonia used 113,500 tons of shale. Two years later the Tallin power station was converted to burn shale, and by 1939 it reached a capacity of 22 MW. In 1928 some 446,000 tons of oil shale were extracted; by 1938 output had reached 1,472,000 tons, and it rose to 1,886,000 in 1940.

At the same time, production of oil shale from the Leningrad and Volga fields was of considerable importance to local industries, but total output in the USSR remained well below that of the Estonian fields. In 1934 the USSR produced 206,000 tons; this rose to 417,000 in 1935, 467,900 in 1936, 505,000 in 1937, 562,100 in 1938, and 736,000 in 1940.

The nonaggression pact signed between the USSR and Nazi Germany in 1939 included a secret protocol assigning the Baltic States to the Soviet sphere of influence; in the summer of 1940 they were annexed by the USSR. Considerable damage was suffered by the shale industry during the war years (1941-45), when all the mines and shale-processing factories were put out of commission.

In 1945 the Soviet government passed a decree on "the reconstruction and development of the shale industry in the Estonian SSR and Leningrad oblast." The main objective was to ensure a supply of manufactured gas to the city of Leningrad. By 1950 reconstruction had been completed and several new mines had been built. In 1948 a factory for manufacturing gas from shales went into production in Kokhtla-Yarve, from which gas was transmitted 208 km. by pipeline to Leningrad. In February 1953 a second pipeline was commissioned to transport gas 140 km. to Tallin.[2]

Since Leningrad and the Baltic States have no local resources of oil, gas, or coal, it was decided immediately after the war to develop shale reserves as rapidly as possible. Although output in

1945 was less than half of that of prewar years, by 1950 the USSR was producing some 4.7 million tons; over double that amount, 10.8 million tons, was extracted in 1955, making the Soviet Union the major shale-producing country in the world. This rapid rate of increase has continued, with 21.3 million tons being produced in 1965 and 26.1 million tons in 1971.[3]

PRESENT AND FUTURE DEVELOPMENT

Location of Reserves

The world's resources of oil shale are as yet little explored, and any attempt to compare the reserves of individual countries is even more difficult because the oil yield per ton of shale varies considerably from one deposit to another. Researchers in the United States have concluded that North America has considerably greater known reserves (recoverable under present circumstances) than Europe and Asia combined.[4] This is not disputed by Soviet experts and is confirmed by papers presented at such international conferences as the symposium on oil shale resources held by the UN in Tallin in 1968.

Estimates of total reserves, however, tend to change the picture in favor of the USSR, which has considerable potential in areas not yet subjected to intensive exploration. One 1968 estimate puts total geological reserves of oil shale in the USSR at 662,500 million tons.[5] Geological reserves of oil shale in the USSR are, however, more often given as 156,000 million tons, which could prove to be a considerable underestimation, since it is based on a survey made in 1956. The exploration of shale reserves has been given low priority because of the small share of this fuel in the Soviet energy balance and the rapid growth in output of more economical fuels. Total balance reserves are estimated at 54,000 million tons. Balance reserves in categories $A + B + C_1 + C_2$ as of 1968 totalled 16,500 million tons, of which some 3,500 million had been proved to categories $A + B$, 3,000 million to category C_1, and 10,000 million to category C_2.

Of this total, 52 percent is situated in Estonia (8,500 million tons), and a further 19 percent (3,000 million tons) is in the neighboring Leningrad oblast.[6] There is also a large basin in the Volga area, mainly in Kuibyshev oblast, that has about 16 percent (2,600 million tons) of the shale reserves. The distribution of known resources is given in detail in Tables 6.1 and 6.2 and is shown on Map 6.1.

The Baltic basin. Situated mainly in Estonia, but stretches into Leningrad, Novgorod, and Pskov oblasts, covering an area of about

198

TABLE 6.1

Reserves of Oil Shale, by Deposit
(in millions of tons)

Field or Basin	Total Geological Reserves as of 1956	Industrial Reserves $(A + B + C_1)$ as of 1965	Technically Possible Annual Output in Near Future
USSR	156,000	6,180	74
Estonian	14,477	3,501	41
Leningrad	5,953	1,149	18
Kashpirskoe	2,126	702	6
Obshchii Syrt	4,060	376	9
Others	129,384	452	—
(of which in the north-eastern Siberian platform)	111,800	—	—

Source: Melnikov, N.V., Mineralnoe toplivo, Moscow 1971: 24.

60,000 sq. km. Geological reserves are over 20,000 million tons, of which the balance reserves (recoverable under present conditions) amount to 11,700 million tons. Layers are almost horizontal, lie mostly at depths of less than 90 m., and vary greatly in thickness, the maximum being about 22 m. and the average about 3 m. Shales are extracted at depths from 8 to 45 m. Exploitation of the Estonian fields began in 1916, and by 1966 they had yielded 190 million tons of oil shale, mainly from mines. In the next six years (by the beginning of 1973) a further 103 million tons were extracted. Three opencast workings, with a total capacity of 9.5 million tons a year, have been constructed in the north of the basin to supply the fuel for the Baltic power station. Estonian shale is of high quality, the organic content, kerogen, having on average 77 percent carbon, 10 percent hydrogen, 10 percent oxygen, 2 percent sulphur, .3 percent nitrogen, and .7 percent chlorine; the calorific value of kerogen is around 9,000 k.cal/kg. The ash content of the shale varies between 40 percent and 55 percent, the carbon dioxide between 22 percent and 23 percent, and the moisture content from 9 percent to 13 percent. The calorific value of "run-of-

Legend

Basins				
A	Baltic		7	Kubinskoe
B	Carpathian		8	Baisunskoe
C	Belorussian		9	Ravanskoe
D	Volga		10	Garautskoe
E	Pechora-Vychegda		11	Terekhlitausskoe
F	Olenek		12	Kenderlykskoe
			13	Dmitrievskoe
			14	Budagovskoe, Alyuiskoe
Fields			15	Khakhareiskoe
1	Boltyshskoe		16	Gusino-ozerskoe
2	Novodmitrovskoe		17	Romanovskoe
3	Rionskoe		18	Yumurchenskoe
4	Kotainskoe		19	Turginskoe
5	Norarevikskoe		20	Kharanorskoe
6	Dzhengigaiskoe		21	Omolonskoe

MAP 6.1

Oil Shale Reserves

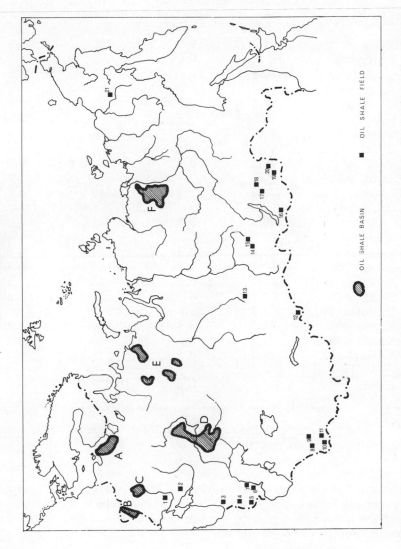

Source: Razrabotka i ispolzovanie zapasov goryuchikh slantsev, Tallin 1970: 26.

TABLE 6.2

Distribution of Known Oil Shale Reserves Recoverable under Present Conditions

Area	$A + B + C_1$ Reserves (in millions of tons)	(in percentage)	$A + B + C_1 + C_2$ Reserves (in millions of tons)	(in percentage)
RSFSR	2,567.4	39.0	7,698.3	46.3
Leningrad oblast	1,120.1	17.0	3,075.7	18.5
Kostroma oblast	6.1	0.1	42.3	0.2
Kuibyshev oblast	728.7	11.1	2,623.9	15.8
Ulyanovsk oblast	48.2	0.7	160.6	1.0
Saratovsk oblast	143.8	2.2	146.0	0.9
Orenburg oblast	376.0	5.7	833.3	5.0
Kemerovo oblast	41.8	0.6	131.8	0.8
Irkutsk oblast	90.5	1.4	90.5	0.5
Bashkir ASSR	10.7	0.2	42.7	0.3
Komi ASSR	1.6	—	551.6	3.3
Estonian SSR	3,931.8	59.8	8,599.2	51.8
Kazakh SSR	54.0	0.8	217.7	1.3
Kirgiz SSR	21.9	0.3	65.2	0.4
Tadzhik SSR	3.9	0.1	30.0	0.2
USSR	6,579.0	100	16,610.4	100

Source: Energeticheskie resursy SSSR: Toplivno-energetic-heskie resursy, Moscow 1968: 604-605.

the-mine" shale ranges from 2,000 to 3,000 k.cal/kg. The Leningrad fields have been exploited since 1932 and the yield is much smaller, only 5 million tons a year by the beginning of the 1970s. Shales are mined at depths of 60 to 100 m.[7]

The Volga basin. Extends through Kuibyshev, Saratov, Orenburg, and Ulyanovsk oblasts and part of the Tatar ASSR. Geological reserves

202

are put at 5,000 million tons, of which some 3,800 million (23 percent of the total known Soviet reserves) are recoverable under present conditions. There are nine fields, of which only the Kashpirskoe, 12 to 15 km. south of Syzran, is being exploited. With the advent of natural gas, two others were closed as being uneconomical. The quality of Volga basin shale fluctuates greatly but averages well below that of Baltic basin shale. Mining is at depths of 60 to 230 m.[8]

In the 1960s significant deposits were discovered in Belorussia and in the Cherkassy and Kirovograd oblasts of the Ukraine. The Boltyshskoe field in the Ukraine covers an area of 260 sq. km., with shales lying at depths of 90 to 280 m. Total reserves are estimated at 3,400 million tons with a calorific value of 2,200 to 2,700 k.cal/kg.

Belorussia. The shale reserves are not yet fully explored and appear to be of comparatively poor quality. In 1972 new deposits were discovered, covering an area of over 20,000 sq. km.; these new deposits are calculated to amount to about 11,000 million tons geological reserves. The most suitable areas for exploitation are around Lyuban and Zhitkovichi, to the south of Minsk.[9]

About fifty shale fields are known in the USSR. In addition to those mentioned above, there are shale deposits in the Kirov and Kostroma oblasts; in the Chuvash, Mordovian, and Komi autonomous republics; and in the Urals, the northern Caucasus, and Siberia; there are also known reserves in Kazakhstan and in the republics of Central Asia.[10]

They are not well explored, and only in Kazakhstan have shales actually been extracted in small quantities. The Kenderlyk-Zaisan area is the richest in shale, with some 2,000 to 3,000 million tons geological reserves. Output of shale in Kazakhstan was never great, reaching 17,000 tons in 1950 and ceasing altogether in 1956.[11]

Production

Oil shale has been commercially exploited in many countries, including France, Scotland, Germany, Sweden, South Africa, China, Canada, and the United States. The USSR consumes considerably more shale than any other country, particularly in Estonia, although since the oil yield from shales varies between 10 and 130 gallons per ton it is difficult to make exact comparisons.

According to a February 1973 speech made by B. E. Bratchenko, the USSR Minister of the Coal Ministry, Soviet production of oil shale in that year was to exceed 30 million tons.[12]

There are three important areas of production in the Soviet Union: Estonia, Leningrad oblast, and Kuibyshev oblast. Of these

Estonia has consistently produced more than the other areas combined, and accounts for some 80 percent of the total output. (See Table 6.3 and Figure 6.1.)

Shale is extracted from high-capacity mines, where highly mechanized methods may be applied. In Estonia there are nine mines and three opencasts, the latter accounting for some 40 percent of shale output. Number nine shale mine was commissioned in January 1973 with a yearly capacity of 3.4 million tons; this was to be increased to 5 million tons in 1974 when the second stage went into operation. Its estimated capacity is 9.2 million tons of shale mass a year. By 1972 mechanization of output in Estonia had reached 76 percent, and mechanical shale dressing 39 percent. Labor productivity had risen to 160 tons per man per month in mines and to 500 tons in opencast extraction. Cost per ton was 2 rubles 50 kopeks.[13]

In Leningrad oblast there are four mines; the largest, near Slantsy, is expected to reach its rated capacity of 5 million tons a year by 1975. By the beginning of the 1970s there were only two mines in operation in Kuibyshev oblast, both at the Kashpirskoe field.

In 1970 the average daily output per mine was 4,530 tons for the entire USSR, 5,375 tons in Estonia, 3,773 tons in Leningrad oblast, and 2,242 tons at the Kashpirskoe field. This shows a considerable increase since 1950, when the average for the USSR was only 790 tons, and even since 1965, when the average was 3,560 tons. The average output in shale mines has generally been over twice the output in coal mines.[14]

Extraction is mainly by shot-firing. Loading has been mechanized almost completely at development faces and by 1966 was 37 percent mechanized in longwall sections. Shale is transported from the mines by conveyors, by electric locomotives, or by trucks with rubber tires. In 1968 the equipment in the Baltic basin mines alone included the following: 130 shale-cutting machines, 192 mechanical loaders, about 600 conveyors, 256 electric locomotives, 40 excavators, and 67 bulldozers. Labor productivity has been considerably improved; average output per man per month has increased from under 40 tons in 1950 to over 140 tons in 1970. The average cost per ton in 1969 was 3 rubles 11 kopeks, being cheapest in Estonian opencast extraction, where the cost dropped to 1 ruble 82 kopeks.

The extraction cost of shale per ton of conventional fuel is approximately the same as for coal, although opencast shale is about 40 percent cheaper. Shale is therefore only economically competitive when it is both produced and consumed in an area deficient in other forms of fuel.[15]

TABLE 6.3

Oil Shale Production, 1945–72
(in millions of tons)

Year	USSR	RSFSR Volga Fields (Kuibyshev oblast only, since 1957)	RSFSR Leningrad Oblast	Estonia	Total Output for USSR in millions of Conventional Tons
1945	1.4	0.5	0.0	0.9	0.4
1950	4.7	0.8	0.3	3.5	1.3
1955	10.8	1.7	2.1	7.0	3.3
1960	14.1	1.4	3.5	9.2	4.8
1965	21.3	1.3	4.1	15.8	7.4
1966	21.4	1.1	4.2	16.1	7.5
1967	21.6	1.2	4.3	16.1	7.5
1968	21.9	1.1	4.3	16.4	7.6
1969	23.0	1.2	4.3	17.5	8.0
1970	24.3	5.4		18.9	8.8
1971	26.1	5.3		20.8	9.5
1972*	29.3	5.5		23.7	9.9
1973*	31.0	6.0		25.0	NA
1975†	37.0	1.2	5.8	30.0	NA

NA: Not available

*Approximately.
†Planned.

Sources: Melnikov, N. V., Mineralnoe toplivo, Moscow 1971: 141; Narodnoe khozyaistvo SSSR v 1970 g., Moscow 1971: 183, 189; SSSR v tsifrakh v 1971 godu, Moscow 1972: 87; Sovetskaya Estoniya, 14 December 1972: 2; Ugol no. 1, January 1974: 5.

FIGURE 6.1

Oil Shale Production in USSR, 1950-1973
and Planned for 1975

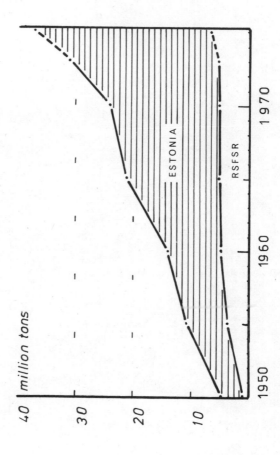

Source: Table 6.3.

Utilization

Because of its low calorific value, oil shale is regarded as a fuel of mainly local significance. By the 1970s some 62 percent of the Soviet shale output was consumed to produce electric power, and the remaining 38 percent was used for other industrial purposes.[16]

Energy shale should have a calorific value of 2,600 to 3,100 k.cal/kg., while shale used for industrial processing varies from 3,100 to 3,500 k.cal/kg. Since sorting and cleaning by hand is expensive and inefficient, the process has been mechanized to a large extent. Work on the first plant began in 1958, and by the 1970s there were five shale cleaning works in Estonia with a total capacity of some 15 million tons.[17]

The main power stations in Estonia are fueled by oil shale. In 1959 the Baltic power station near Narva went into production, reaching a capacity of 1,900 MW by the 1970s; in 1973 a second large station, the Estonian, of 2,400 MW capacity, was being completed, having first gone on load in 1969.[18]

There are two shale-fired power stations in Slantsy, in Leningrad oblast, and one in Syzran, in Kuibyshev oblast. In 1955 the Saratov power station, which had formerly burned shale, was converted to natural gas. By 1970 shales were used to generate about 1.4 percent of Soviet electric power, mainly due to the large Baltic and Estonian stations. In 1967 the Baltic station generated 7,500 million kWh. at a cost of .75 kopeks per kWh, which was considerably cheaper than the average for the Northwest. It also supplied hot water for heating factories and homes in Narva.[19]

The burning of oil shales in power stations gives shale ash, an important by-product used in the manufacture of high-grade cement, as well as bricks and other building materials. In Estonia and Leningrad oblast, large shale-processing combines have been built to supply domestic gas, boiler fuel, gasoline, sulphur, creosote, coke, bitumen, tanning agents, benzene, toluene, and other useful products. At the processing works in Kuibyshev oblast medicinal components are prepared, as well as chemicals for the dehydration and desulphuration of oil. It is cheaper to make such products from oil shale in the area of its extraction than it is to use oil, gas, or coal that has had to be transported over long distances.[20]

Estonian scientists have recently discovered how to make a synthetic cleaning fluid from shale that is 50 percent more efficient than normal soap at only a third of the cost. They are also manufacturing useful insecticides and substances that help to combat soil erosion. Over 4 million tons of ash remains each year from the shale consumed in Estonian power stations. Not all of it is used for cement or building aggregates. Large quantities are also fed into acid soils in the Northwest region for chalking purposes.[21]

In March 1973 an article in Pravda emphasized the importance of kerogen, the organic material obtained from oil shale. Researchers in the All-Union Institute of Petrochemical Processes predict that some 80 to 100,000 tons a year will soon be needed by the various branches of the chemical industry. Certain substances can only be extracted from kerogen and cannot be obtained from petroleum, natural gas, or coal. The author criticized the Soviet Petrochemical Industry for not expanding production of kerogen and intensifying research into its qualities quickly enough to meet this growing demand.[22]

Future Production

By 1975 the capacity of the oil shale mines in the USSR is expected to be 37 million tons, of which Estonian mines will contribute 30 million tons, Leningrad mines 5.8 million tons, and Kuibyshev mines 1.2 million tons. Two further opencasts are to be completed in Estonia, bringing the proportion of opencast extraction to 36 percent in 1975. Mechanization will be intensified in the underground mines, especially in loading at the work face, which should reach 93 percent by 1975. By that year the average daily output per mine is planned to be 7,230 tons and labor productivity 180 to 190 tons per man per month. Judging by past and present rates of progress such figures may well be attained.[23]

The rapid expansion of oil shale production in the Baltic Basin is likely to continue, reaching an annual output of about 50 million tons in the near future. In the Baltic area a conventional ton of shale fuel costs 50 percent less than imported coal from the Donbass, Kuzbass, or Moscow basins, 30 percent less than fuel oil, and 42 percent less than gas from the nearest fields. Further developments in shale utilization will include improving the systems that combine the generating of energy with industrial processing to make the useful products mentioned above.[24]

The direct consumption of shale to produce electricity is complicated by many factors, including high ash content and air pollution by sulphurous gases. At present only comparatively high quality shale, over 2,200 k.cal/kg., is used. However, various methods of combining the chemical processing of lower quality shale with the generating of electricity are being developed at the Krzhizhanovskii Energy Institute and other research centers in the Northwest.[25]

Despite the completion of pipelines bringing cheap natural gas to the Northwest, the Soviet shale industry is expected to continue to expand its output throughout the 1970s, making an extremely valuable contribution to the energy balance of this important economic region.

NOTES

1. Volkov, T. M., et al., Razrabotka i ispolzovanie zapasov goryuchikh slantsev (The development and utilization of oil shale resources), Tallin 1970: 622.

2. "Slantsevaya promyshlennost" (Oil shale industry), Bolshaya Sovetskaya Entsiklopediya, Moscow 1949-57, Vol. 39; Ponomarev, B. N., et al., Istoriya SSSR (History of the USSR), Moscow 1967, vols. 7, 8; Estoniya (Estonia), Moscow 1967: 90-91; Kotov, M. A., et al., "Sostoyanie i perspektivy razvitiya slantsepererabatyvayushchei promyshlennosti SSSR" (The state of and prospects for development in the oil shale processing industry of the USSR), in Volkov, op. cit., 231.

3. SSSR v tsifrakh v 1971 godu (USSR in statistics for 1971), Moscow 1972: 86-87.

4. Resources and Man, a study by the U.S. National Academy of Sciences and National Research Council, San Francisco 1967: 198-200.

5. Volkov, op. cit.; 26-41; Sidorenko, A. V., et al. 50 let Sovetskoi geologii (50 years of Soviet geology), Moscow 1968: 362.

6. Melnikov, N. V., Mineralnoe toplivo (Mineral fuel), 2d. ed., Moscow 1971: 140; Rostovtsev, M. I., Runova, T. G. Dobyvayushchaya promyshlennost SSSR, Moscow 1972: 70-71.

7. Energeticheskie resursy SSSR: Toplivno-energeticheskie resursy (Energy resources of the USSR; fuel resources), Moscow 1968: 603-12.

8. Ibid.: 612-14.

9. Pravda, 18 November 1972: 2.

10. Energeticheskie resursy, op. cit.: 614; "Geologiya i perspektivy osvoeniya mestorozhednii goryuchikh slantsev SSSR" (The geology and prospects of developing deposits of combustible shales in the USSR), in Volkov, op. cit., 25-35.

11. Kushev, G. L., Ugolnye bogatstva Kazakhstana (Coal riches of Kazakhstan), Alma Ata 1971: 52-53; Melnikov, op. cit.: 141.

12. BBC Monitoring Service, USSR Weekly Economic Report, 2 March 1973.

13. Sovetskaya Estoniya, 29 December 1972; BBC Monitoring Service, USSR Weekly Economic Report, 2 and 16 February, 1973; Ugol no. 12, December 1972: 28; Sovetskii Soyuz no. 5, May 1973: 31.

14. Melnikov, op. cit.: 141.

15. Ibid.: 142; Energeticheskie resursy, op. cit.: 615-17; Petrov, A. P., "Perspektivy razvitiya promyshlennosti goryuchikh slantsev v SSSR" (Prospects for the development of the oil shale industry in the USSR), in Volkov, op. cit.: 95.

16. Melnikov, op. cit.: 143.

17. Makovskii, Yu. A., "Metody obogashcheniya goryuchikh slantsev" (Methods of dressing oil shale), in Volkov, op. cit.: 142-49.

18. Sotsialisticheskaya industriya, 21 July 1970: 2; BBC Monitoring Service, USSR Weekly Economic Report, 22 June 1973.

19. Energeticheskie resursy, op. cit.: 612, 621; Volkov, op. cit.: 374, 570.

20. Ibid.: 623; Ravich, M. B., Toplivo i efektivnost ego ispolzovaniya (Fuel and its efficient utilization), Moscow 1971: 180-83.

21. Estoniya, Moscow 1967: 90-95.

22. Pravda, 23 March 1973.

23. Volkov, op. cit.: 96-98.

24. Volfson, G.B., "Ekonomicheskaya tselesoobraznost razvitiya dobychi goryuchikh slantsev v SSSR" (The economic advisability of developing oil shale production in the USSR), in Volkov, op. cit.: 156-57.

25. Tyagunov, B.I., et al., "Energo-tekhnologichcskoe ispolzovanie pribaltiiskikh slantsev" (Power and technological utilization of Baltic shale), Volkov, op. cit.: 394-400.

The electricity industry, which is both a primary producer of
energy and a major consumer of mineral fuels, deserves to be discussed separately. While over 80 percent of Soviet electric power is
produced in thermal stations burning coal, petroleum, gas, peat, or
oil shale, alternative sources such as hydroelectricity and nuclear
energy are making an increasingly significant contribution to the Soviet
energy balance. Electricity transmission lines offer an important
alternative to the more traditional methods of transporting energy,
using roads, rivers, railways, or pipelines.

HISTORICAL BACKGROUND

In 1913 the area now occupied by the USSR produced only 1,945
million kWh of electricity, and by the end of the Civil War in 1921
production had dropped to 520 million kWh. In February 1920 the State
Commission for the Electrification of Russia (GOELRO) was established to assist in the reconstruction of the devastated economy over
a ten to fifteen year period by the building of twenty large thermal
and ten hydroelectric power stations with a total installed capacity of
1,500 megawatts (MW). Much emphasis was placed on the utilization
of cheap local fuels and the formation of electricity grids. Peat,
lignite from the Moscow Basin, anthracite fines from the Donbass,
oil shale, and even timber-processing waste began to be used in thermal

Some recent data have been added to this chapter from the BBC
monitoring service, USSR Weekly Economic Reports, 1972-73.

plants with reasonable success, and the utilization of local fuels has become a permanent part of Soviet power policy. Early construction methods were fairly primitive, with manual excavation and concrete poured from wheelbarrows, and work stopped almost completely in winter. Since labor productivity was very low, power units being generally assembled on site, it normally took five to six years to complete these first thermal plants. Nonetheless, by the beginning of the first Five Year Plan the total capacity of Soviet power stations was 2,296 MW, and output for 1929 had reached 6,200 million kWh. So much effort was put into this sector of the economy that by 1935 forty (instead of the planned thirty) power stations had been constructed.[1]

Because of the threat of German invasion, particular attention was paid to developing the eastern areas of the USSR. While in the country as a whole in the period from 1913 to 1940 electricity production increased 23.7 times to 48,309 million kWh, it is claimed, for instance, that a farsighted government effort multiplied output in Kazakhstan 486 times and in Central Asia 84 times.[2] Unfortunately, actual output by 1940 in the eastern areas was less impressive—only 4,500 million kWh, or 9.2 percent was produced east of the Urals in 1940—and the Soviet power industry suffered a catastrophic drop in production when the industrial areas in the west were occupied in 1941.

During the war, however, everything possible was done to expand the number of power stations beyond the Urals, and in 1942 the USSR produced 29,100 million kWh, or about 60 percent of the prewar output, in spite of the loss that had been inflicted. Energy production in the Urals in 1942 was 50 percent more than in 1940, while in the Volga regions and in West Siberia it rose by 33 percent. There was particularly rapid construction of power stations in Sverdlovsk and Chelyabinsk oblasts. In 1943 production for the whole country rose to 32,300 million kWh, 11 percent more than in 1942, and by 1944 it had reached 39,200 million kWh.[3]

It was not until 1946 that the electricity output in the USSR surpassed the 1940 level, but expansion since then has been very rapid indeed.[4] (See Table 7.1.) The war had made it necessary to speed up construction work on new plants. Mechanization of installation and assembly, and prefabricated steel reinforcements, helped to reduce the time needed on the site to such an extent that by the 1970s thermal plants of over a million kw capacity could be constructed in less than a year.

In order to keep this rapid growth in perspective, however, it is worth bearing in mind that the USSR is far from drawing level with the United States in production of electricity. (See Table 7.2 and Figure 7.1.) In 1971, when the United States generated 1,820 million

TABLE 7.1

Production of Electricity in USSR, 1940-80

Year	Capacity (in thousands of MW)	Production (in millions of MWh)
1940	11.2	48.3
1945	11.1	43.3
1950	19.6	91.2
1955	37.2	170.2
1960	66.7	292.3
1965	115.0	506.7
1966	123.0	544.6
1967	131.7	587.7
1968	142.5	638.7
1969	153.8	689.1
1970	166.2	740.4
1971	175.4	800.0
1972	186.0	858.0 (850*)
1973	NA†	913*
1974	207.0*	985*
1975	230.0*	1,065*
1980	330.0*	1,800*

*Planned.
†NA: not available.

Sources: Narodnoe khozyaistvo SSSR 1922-1972, Moscow 1972: 158; Pravda, 22 December 1971: 3; Pravda, 30 January 1973: 1; Ekonomicheskaya gazeta no. 4, January 1973; Pravda, 26 January 1974: 2; Novoe vremya, no. 9, March 1974: 14.

MWh, the Soviet Union generated only 800 million MWh, or 44 percent of this. It is also significant that while 775 million MWh (43 percent of total output) was used in American industry, a much higher production of the Soviet total, 522 million MWh (65 percent), was consumed for industrial needs. This reflects the greater domestic consumption and higher material standard of living in the United States. Indeed, while the USSR is second only to the United States in total output, on a production per capita basis it falls considerably below most industrial nations of the West. In 1970, for example, the USSR generated 3,052 kWh per inhabitant, compared to 8,423 kWh in the United States.

213

TABLE 7.2

Comparative Production of Electricity in USSR
and United States, 1940-80
(in millions of MWh)

Year	USSR	United States
1940	48	180
1946	49	269
1950	91	389
1955	170	629
1960	292	842
1965	507	1,158
1970	740	1,750
1971	800	1,820
1975	1,065†	2,194*†
1980	1,800†	3,086*†

*Excludes generation by industrial facilities and railways having generating facilities of their own, which could amount to an additional 150 million MWh.
†Planned.

Sources: UN Statistical Yearbook 1970, New York 1971: 376; SSSR v tsifrakh v 1971 godu, Moscow 1972: 60-61; Statistical Abstract of the United States, 1971, Washington 1971: 497. UN Statistical Yearbook 1972, New York 1973: 373; Novoe vremya, no. 9, March 1974: 14.

The USSR is not likely to surpass the 1970 production of electricity in the United States until 1980, and even then it will probably be generating little more than half as much electricity as the United States by that year.[5]

The production of electricity in the USSR has been developed according to four basic principles: the construction of large regional power stations using cheap fuel and water power; the combined production of heat and power in single plants for use in towns and industrial centers; the building of hydroelectric schemes in such a way as to improve river transport and land irrigation; and the formation of a series of power grids that will eventually be linked into a single system for the whole USSR.[6]

Combining the production of heat and power reduces fuel consumption to 50 to 60 percent of the amount required to generate the same electricity and heat at separate plants, although the initial cost of such a plant is higher than that of a condensing turbine plant. Since 1931 increasing importance has been placed on the building of large heat and power plants, especially in the Moscow, Leningrad, Urals, and Donbass conurbations. In 1970 such plants generated 12 percent of the total output. The largest-ever heat and power plants were being built in Moscow and Leningrad in 1973, using 250 MW turbines.

By far the greatest proportion of power production (61 percent in 1970) is still from the condensing turbine plants, which can be sited either at the source of fuel or in the area of consumption, provided that there is an adequate supply of cooling water. Since 1955 the capacities of these plants has increased greatly. The Konakovo regional power station, which functions in the Central Economic Region on natural gas from the North Caucasus, has a capacity of 2,400 MW. Other stations of similar capacity are the Nazarovo and Irsha-Borodino, in the Kansk-Achinsk coal basin; the Tom-Usinsk and Belovo, using Kuzbass coal; the Ermak, using Ekibastuz coal; the Lugansk, Zmiev, Mironov, Pridneprovsk, and Krivoi Rog, using Donbass coal; and the Estonian and Baltic, using Estonian oil shale.[7]

In 1973 there were forty-five thermal power stations with a capacity of over 1,000 MW, fourteen of which were over 2,000 MW. The Kostroma station on the Volga has a planned capacity of 4,800 MW. A power unit of 300 MW has become standard equipment for large thermal power stations. In June 1972 it was decided to add a ninth and tenth unit of this size to the Krivoi Rog station, raising its capacity to 3,000 MW in November 1973, making it the largest thermal plant in the world. The Uglegorsk station, which will utilize cheap Donbass coal, has a planned capacity of 3,600 MW from four units of 300 MW and three of 600 MW. The first two blocks were installed in 1972.[8]

A significant proportion of the electricity generated in the Soviet Union has always been supplied by hydroelectric stations; in 1972 this was as much as 14.4 percent, in comparison to 84.7 percent generated in thermal stations and .9 percent in nuclear stations. In Tsarist Russia there were only three hydroelectric power stations, with a total capacity of 16 MW, but in the Soviet period much effort and publicity has been allotted to this sphere of construction, and by the 1970s hydroelectric power capacity was over 30,000 MW, producing over 120 million MWh.

Hydroelectric projects, involving as they do the spectacular damming of powerful rivers, tend to capture the imagination of the public much more than even the largest thermal power stations. This is often reflected in the names chosen by the authorities. In 1957 the "V.I. Lenin Volga Hydroelectric Power Station" went into operation,

FIGURE 7.1

Comparative Production of Electricity in USSR
and United States, 1940-1975

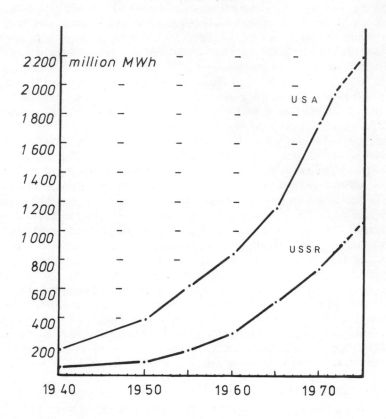

Source: Table 7.2.

reaching a capacity of 2,300 MW, and in 1961 there was even the "22nd CPSU Congress Hydroelectric Power Station" (also on the Volga) with a capacity of 2,500 MW. In the same year the Bratsk Hydroelectric Power Station, of 4,000 MW capacity, went into operation on the Angara river. The largest hydroelectric station built in the 1960s was the Krasnoyarsk, on the Enisei, which began production in 1967, reaching a total capacity of 6,000 MW in 1971. Even larger is the Sayano-Shushensk station, also on the Enisei, with a planned capacity of 6,400 MW.[9]

PRESENT AND FUTURE DEVELOPMENT

The fuels from which over 80 percent of Soviet electricity is generated are discussed in the previous chapters. The other important factors in the energy balance are hydroelectric and atomic power.

The USSR is estimated to have 12 percent of the world's resources of water power, with a potential annual electricity production of 3,338 million MWh, of which it may prove economically feasible to exploit 1,095 million MWh. About 70 percent of these resources are situated in Siberia and the Far East, including the vast Enisei, Lena, Ob, Angara, and Irtysh rivers. Kazakhstan and Central Asia contain about 15 percent, with such rapidly flowing rivers as the Naryn, Amu Darya, Syr Darya, and Chirchik. The Caucasus area has 8 percent; in the rest of the European part of the USSR, with only 7 percent, the most important sources of hydroelectric power are the Volga and Dnepr rivers. Since the utilization of hydroelectric resources in the USSR is still only a fraction of what may eventually be possible, further large-scale construction of hydroelectric schemes is planned.

Although the rivers Lena and Enisei are potentially the largest contributors of hydroelectric power in the USSR, there are no plans to utilize the energy of the Lena before 1990, and the Enisei is producing only 57 percent of its possible output (see Table 7.3). The extension of the unified electricity grid to East Siberia and the Far East would be one method of using this potential power. The abundant supplies of hydroelectricity could also be used to exploit the untapped resources of valuable minerals in the east of the USSR, forming new industrial complexes.[10]

Experiments have been conducted for several years on utilizing tidal power resources in the USSR, which have been calculated at 210 million MWh, about one-sixth of the world's total. The most suitable places on the Soviet coastline for constructing tidal power stations are in the Far East (Kolpakova Estuary, the Shantar Islands, Nogaevo Bay, and the Shelekhova Gulf), where there is little need for expensive additional energy. The White Sea, however, has a potential of 40

TABLE 7.3

Hydroelectric Power

Area	Economically Viable Resources			Degree of Utilization (in percentage)		
	(in millions of MWh)	(in millions of kW)	(in Percentage)	1965	1970	Designed Level
USSR	1095	120	100.0	2.4	3.4	7.8
European USSR	192	18	17.5	8.5	10.4	16.2
Eastern Siberia	350	40	32.0	2.2	4.7	12.5
Enisei River	288	32	26.0	2.2	4.7	12.5
Far East	294	34	26.9	—	0.1	0.7
Lena River	235	27	21.4	—	—	—

Source: Rylskii, V. A., Ekonomika mezhraionnykh elektro-energeticheskikh svyazei v SSSR, Moscow 1972: 122.

million MWh, which might be included in the European energy system were tidal power stations to prove economically feasible. In 1962 it was decided to build the experimental Kislogub station on the coast to the northwest of Murmansk, by prefabricating sections and floating them into position. This first station, with a capacity of 800 MW, was successful, but too costly to bring immediate approval for the more adventurous Lumbovskaya, Mezen Bay, and White Sea schemes. In 1973, however, after the Kislogub station had been in operation for six years, research work was continuing for new stations in Mezen Bay (6,000 MW) and Tugur Bay (10,000 MW) on the Sea of Okhotsk. Also being planned was a tidal power station for Kara-Bogaz-Gol in the Caspian Sea, which is to be assembled in Baku and towed across. For the Kamchatka tidal power station (25,000 MW) it will be necessary to construct a dam across the Penzhinskaya Guba, from Mys Srednii on the west coast to Mys Vodopadnyi on the east coast.[11]

In view of its vast reserves of fossil fuels and its large hydroelectric potential, the Soviet Union would appear to have less urgent need to develop nuclear energy than other industrial nations, and indeed at the beginning of the 1970s atomic energy capacity in the USSR was well below that of Britain and the United States. There are, however, strong arguments in favor of an accelerated development program. Because of the isolation and inhospitable climate of some areas with rich mineral deposits, difficulty has been experienced in supplying them with adequate power by conventional means. Atomic power stations do not rely on roads or railways, since small quantities of fuel are sufficient for long periods of production. This is also an advantage in the many industrial regions of the European USSR that have not enough local energy resources. The economic partners of the USSR in Eastern Europe cannot satisfy their own energy requirements, and nuclear energy will become an increasingly more rational alternative to the long-distance transport of fuels from beyond the Urals.

On 27 June 1954, the USSR completed its first nuclear power station in Obninsk, southwest of Moscow, with a capacity of 5 MW. The first large plant, of 100 MW capacity, began production in 1958 somewhere in Siberia, probably to produce plutonium for nuclear weapons. By 1970 its capacity had been increased to over 600 MW. In 1965 a boiling-water reactor was started five km. from Melekess, rising from the original capacity of 50 MW to 200 MW by 1969.

On the banks of the Don, about 45 km. from Voronezh, is the Novovoronezh atomic power station of the light water reactor type. In December 1969, when Unit 2 went on line, Unit 1, which had been in operation since September 1964, had to be closed for over a year to repair the reactor-vessel thermal shield. By April 1973, however, the Unit 4 had been installed and the station's aggregate capacity had reached 1,500 MW; it is expected to rise to 2,500 MW when the Unit 5 goes on line in 1975.

In April 1964, in the Urals east of Sverdlovsk, the first super-heater reactor of the Beloyarsk atomic power station was started at 100 MW, being joined by the second in December 1967 at 200 MW. In 1972 a fast neutron reactor was under construction at Beloyarsk.

The Shevchenko breeder reactor atomic power station on the east coast of the Caspian Sea has a capacity of 350 MW and is designed to supply up to 120,000 cu. m. of fresh water daily by desalination, for the population of the arid Mangyshlak peninsula. It went on load in July 1973. (Reports in the Western press in February 1974 that the Shevchenko station had suffered a serious accident were sharply denied in the Soviet media.)

In 1972 the first of two 440 MW reactors was installed at the Kola atomic power station south of Murmansk, and it began to supply power for the Kola energy grid in July 1973. The second unit was expected to go on load in December 1974.

The smaller 48 MW reactor being built at the Bilibino gold-mining center in the Chukotka peninsula in the Soviet Far East is expected to go on load in 1975. In April 1973 the assembly of the first turbine (12 MW) was completed. The plants will also supply the homes and factories of Bilibino with heating and hot water.

A large increase in nuclear energy, 7,200 MW, is planned for the ninth Five Year Plan (1971-75), and it is hoped that the cost will fall to that of conventional thermal power stations. The Armenian station at Oktomberyan in the Ararat valley is being equipped with two 440 MW reactors and has an initial planned capacity of 815 MW. By the end of 1971 work had begun on the Leningrad, Kursk, and Smolensk atomic power stations in the RSFSR and on the Chernobyl station in the Ukraine, where two additional atomic power stations are also planned. The Leningrad plant will be the first to be equipped with a reactor of 1,000 MW capacity.

The use of transportable atomic power stations is of great importance in opening up deposits of such valuable minerals as tin, cobalt, gold, and diamonds in inaccessible northern areas that have no local energy resources. The first of these was the experimental TES-3 built in Obninsk in 1961, which led to the construction of the Sever-1 and the Sever-2. Formerly, the only source of power and heating were diesel stations; a station of 1,500 kW would require about 5,000 tons of fuel each season, which had to be shipped by boat and lorry, increasing the cost of fuel up to twenty times. Thirty kg. of uranium 235 would last an atomic station of equal capacity for five years, supplying a settlement of 5,000 inhabitants with both heat and electricity, and only two or three attendants would be needed.

Other smaller generators, such as the Beta-3 and the Efir, using radioactive isotopes, can be left unattended for up to ten years and are ideal for automatic Arctic observation stations.

The nuclear-powered ice-breaker Lenin, built in 1959, has proved to have many advantages over conventional types. Its power system not only saves space on board for the comfort of the crew, but also makes it unnecessary to return to port every three weeks to refuel, and therefore it can keep the northern shipping lanes open for much longer periods than before. The Lenin needs to be refueled only once a year. A second nuclear ice-breaker, Arktika, was reported almost ready for service in December 1973. Western estimates put Soviet nuclear submarine strength in 1973 as high as 111 vessels. This construction program has probably provided useful experience in the technology of the pressurized-water reactors being used at Melekess, Novovoronezh, and Kola.

With plans to raise nuclear energy capacity to 30,000 MW by 1980, (12 percent of the USSR increase in total generating capacity) the Soviet Union cannot be said to be neglecting this sector, in spite of its considerable fossil fuel reserves. In 1972, however, the United States had 31 nuclear power plants in operation (total capacity 15,000 MW), with 86 additional plants being built and 35 more in the planning stages. These 152 plants will have a total capacity of nearly 132,000 MW. Nevertheless, a group of U.S. specialists who visited Soviet atomic energy plants in August 1972 said that they had been impressed by the depth and breadth of Soviet nuclear technology. In addition to pressurized-water reactors, the USSR has developed a safer reactor that does not confine the coolant to a single vessel but divides it among many pressure tubes and that may be refuelled on load. The Leningrad 2,000 MW plant will have twin reactors of this type, and the same type of reactors will be used at Smolensk and Chernobyl. Since Soviet plants are being built in very heavily populated areas, the safety factor is extremely important. The Soviet nuclear energy program does not seem to have been delayed by public protest, as has happened in the United States. In 1974 atomic power stations in the USSR were expected to generate over 16 million MWh, 38 percent more than in 1973.

The USSR is now in a position to sell nuclear technology abroad. Soviet enriched uranium is cheaper than that of the United States, and there is little competition in the small (6 MW to 100 MW) reactor field. There is an obvious market in the East European countries.

In 1966 Soviet specialists completed the equipping of a small atomic power station of 70 MW in Reinsberg, East Germany, and in 1971 the first reactor was installed in the Nord-1 atomic power station on the Baltic coast near Lubmin, the second being planned for 1973. Plants have also been sold to Czechoslovakia and Bulgaria (Kozlodui), while Romania, Poland, and Hungary are planning the construction of Soviet-designed and -equipped nuclear plants after 1975. There is a constant training program in the USSR for nuclear scientists and engineers from all the Comecon countries, a permanent commission on the peaceful utilization of atomic energy having been established in 1960 at the 13th Session of Comecon.

Finland has ordered two Soviet nuclear plants to be built at Loviisa on the south coast, and both are expected to be operational by 1978. The Finnish decision was probably influenced by considerations of fuel supply. Soviet nuclear technology may well find markets even further afield, in view of the rising demand for finished fuel. The French, for example, signed an agreement with the USSR for the first fuel charge for the American-designed reactor at Fessenheim, and this trend may well continue elsewhere.

By the end of the century atomic energy may supply as much as 10 percent of Soviet energy requirements and even more of the requirements of Eastern Europe. By 1980 the total capacity of atomic power stations in Comecon countries will be 40,000 MW, reaching 120,000 to 160,000 MW. by 1990.[12]

The direct conversion of solar energy into electricity, using semiconductors and photosynthesis, is still at the experimental stage, although the successful use of the sun's rays to produce power in the Soviet space program is well known. A project in the Ararat valley of Armenia is planned to generate 1,200 kw., and various schemes in Central Asia have proved feasible on a very small scale.

Wind-powered generators of low capacity have been used for meteorological stations in the Arctic, Kamchatka, and Sakhalin. It is hoped that it will be possible to increase the construction of wind and solar generators of 1 to 15 kw. capacity in the 1970s. By using several thousand of these generators in the inaccessible areas of Kazakhstan, Central Asia, and the Caucasus for irrigation and watering herds from deep wells, there could be considerable saving in labor and expensive transported fuels.

The earth has vast resources of geothermal energy, which several Soviet scientists have advocated utilizing more extensively than at present, especially in Kamchatka, the Kurile Islands, the Chukotka peninsula, Central Asia, and the Caucasus. In July 1973 a survey was completed at Goryachii Plyazh in the Kurile Islands for the Kunashir geothermal station. A geothermal power station with a capacity of 5,000 kw. is already operating in the Pauzhetka valley, Kamchatka, producing electricity at under a tenth of the cost of diesel stations in the same area. The Petropavlovsk geothermal station was being built in 1974; its planned capacity is 300 MW. There are plans to provide heating for over twenty towns in Georgia using geothermal sources. Experiments are continuing for possible production of magnetohydro-dynamic energy (MHD). In summer 1973 an experimental 25 MW magnetohydrodynamic generator was supplying power to the Moscow energy system and MHD installations of higher capacity are now being constructed.

Also of great future importance is research in nuclear fusion, in which significant advances were made by Academician Artsimovich.[13]

Possibly the major problem facing Soviet energy planners is that, while industry and population are concentrated in the west of the USSR, the vast bulk of energy resources of all kinds is in the under-developed eastern regions. In the past electricity has been produced mainly in the areas of greatest consumption, using either local re-sources or transported fuels. The major economic regions for electri-city production have been the Urals, Donets-Dnepr, Volga, and Center, but the contribution of East and West Siberia has been rising rapidly. Production, by republic, since 1940 is shown in Table 7.4.

Roughly three-fifths of the total Soviet production of electricity is generated in the RSFSR, and just under a fifth in the Ukraine. The most important producers among the other republics are Kazakhstan, Uzbekistan, Belorussia, Azerbaidzhan, Estonia, and Georgia. While total output by 1975 was to be 45 percent more than in 1970, output in Tadzhikistan is to rise by 110 percent, in Turkmenia by 90 percent, in Belorussia by 80 percent, and in Georgia and Armenia by 70 per-cent.[14]

TABLE 7.4

Production of Electricity, by Republic, 1940-75
(millions of MWh)

	1940	1950	1960	1965	1970	1972	1975 (planned)
USSR	48.31	91.2	292.3	506.7	740.4	858.0	1,065.0
RSFSR	30.83	63.4	197.0	332.8	470.0	536.0	657.3
Ukrainian SSR	12.41	14.7	53.9	94.6	138.0	158.0	198.5
Kazakh SSR	0.63	2.6	10.5	19.2	34.7	41.3	50.2
Uzbek SSR	0.48	2.7	5.9	11.5	18.3	23.5	31.8
Belorussian SSR	0.51	0.8	3.6	8.4	15.1	21.0	27.0
Azerbaidzhan SSR	1.83	2.9	6.6	10.4	12.0	12.7	15.1
Estonian SSR	0.19	0.4	2.0	7.1	11.6	14.5	16.5
Georgian SSR	0.74	1.4	3.7	6.0	9.0	10.0	15.5
Moldavian SSR	0.02	0.1	0.7	3.1	7.6	9.6	10.0
Lithuanian SSR	0.08	0.2	1.1	3.9	7.4	9.5	11.0
Armenian SSR	0.40	1.0	2.8	2.9	6.1	7.5	11.0
Kirgiz SSR	0.05	0.2	0.9	2.3	3.5	4.1	7.2
Tadzhik SSR	0.06	0.2	1.3	1.6	3.2	3.5	6.8
Latvian SSR	0.25	0.5	1.7	1.5	2.7	2.3	3.6
Turkmen SSR	0.08	0.2	0.8	1.4	1.8	1.8	3.5

Sources: Narodnoe khozyaistvo SSSR v 1969 godu, Moscow 1970: 193; Ekonomicheskaya gazeta no. 6, February 1972: 13; SSSR i soyuznye respubliki v 1972 godu, Moscow 1973.

One sulution to the disparity between the location of industry and energy resources is the development of very high-voltage transmission lines to transmit electricity over long distances from major generating points to areas of maximum consumption. The policy of creating a single electricity grid for the whole of the Soviet Union has the additional advantage of allowing extra current to be fed into each region at its period of peak demand, which in the vast area of the USSR and Eastern Europe combined varies considerably; when it is two o'clock in Moscow it is eleven o'clock in Chukotka.

In 1960 there were 124,000 km. of transmission lines of over 35 kV; by 1970 this had risen to 446,000 km.; and by 1975 it should reach 603,000 km. In 1973 a 750 kV line was being built from Leningrad to Moscow, and a 750 kV line went on load from the Donbass to Lvov.

With improvements in high-voltage technology, lines of up to 800 kV tension have become widely used, and DC transmission of up to 1,500 kV is planned for development in the 1970s. In 1973 experiments were being conducted near Moscow with a section of 1,500 kV line, in preparation for the construction of an Ekibastuz-Center line, which will be 2,500 km. in length. By 1972 a single grid for the European part of the USSR had already been created, incorporating seven regional grids with some 600 power stations, totaling over 110,000 MW. In 1971 they produced 570 million MWh out of the total output of the USSR of 800 million MWh. The next step was to link the united grid of European USSR with the Siberian, Kazakh, and Central Asian grids, linking 640 power stations generating 75 percent of Soviet electrical output. In 1973 there were eleven grids, the first eight of which had already been incorporated into a single system. These grids are described below, and their comparative output of electricity is shown in Table 7.5 and Map 7.1.

1. Center: This grid covers 21 oblasts, including Volgograd and Astrakhan, and incorporates the Konakovo thermal power station of 2,400 MW capacity and the 2,530 MW capacity Volga hydroelectric power station (named for the 22nd Congress of the CPSU). It has several long-distance lines of 500 kV tension.

TABLE 7.5

Production of Electric Power, by Grid, 1972

Grid	Output (in millions of MWh)	1972 (as percentage of 1970)
Center	128.7	111
Northwest	68.5	124
Middle Volga	52.3	109
Urals	128.0	116
South	178.0	117
North Caucasus	20.5	111
Transcaucasia	29.8	111
North Kazakhstan	25.9	123
Siberia	117.6	118
Central Asia	35.4	125

Source: Ekonomicheskaya gazeta no. 4, January 1973: 2.

224

2. Northwest: This covers Belorussia, the Baltic Republics, the Leningrad, Novgorod, and Pskov oblasts, and Karelia. By 1975 the Kola system will be linked to this grid.

3. Middle Volga: This links the networks of Kuibyshev, Saratov, and Penza oblasts and of the Tatar, Chuvash, Mordovian, and Mari autonomous republics; it includes among its power stations the Volga hydroelectric station (named for Lenin) of 2,300 MW capacity and the Saratov hydroelectric station of 1,315 MW capacity.

4. Urals: This covers the Urals area, the Tyumen, Kirov, and Orenburg oblasts, the Udmurt and Bashkir autonomous republics, and neighboring regions of Kazakhstan. Its power is mainly from large thermal stations.

5. South: This has the greatest capacity of all the grids in the European part of the USSR. It covers the Ukraine, Moldavia, and the Rostov oblast. It employs mainly thermal stations, four of which have a capacity of 2,400 MW and two have a 2,300 MW capacity. A transmission line of 750 kV has been constructed over 1,100 km. across the Ukraine from east to west.

6. North Caucasus: This small grid covers mainly the Krasnodar and Stavropol oblasts and extends to the Caucasus mountain range in the south.

7. Transcaucasia: This links the power stations, a high proportion of which are hydroelectric, of Georgia, Armenia, and Azerbaidzhan.

8. North Kazakhstan: This is a comparatively new grid, but it is planned to incorporate thermal stations of up to 5,000 MW capacity based on cheap Ekibastuz coal. Transmission lines of up to 1,500 kV tension will supply its power to the European system.

9. Siberia: This grid incorporates the output of large thermal stations fueled by cheap opencast coal mines, and hydroelectric power stations such as the Krasnoyarsk, of 6,000 MW capacity, and the Bratsk, of 4,100 MW capacity. In 1971 the total capacity of the stations in this grid was 22,500 MW, producing about 100 million MWh.

10. Central Asia: This covers the Uzbek, Turkmen, Kirgiz, and Tadzhik republics and South Kazakhstan. By 1975 the output from the Nurek, Toktogul, and Karakum high-capacity hydroelectric stations will be added. In 1972 work started on the first 300 km. section of a 500 kV transmission line across Kazakhstan, linking the Siberian and Central Asian grids.

11. Far East: In 1972 work was in process to link the power stations of the Amursk and Khabarovsk regions. Three thermal stations on Sakhalin and the Primorsk station may also be incorporated.[15]

In 1959 the Comecon Permanent Commission for Power agreed to construct eight intergrid transmission lines totaling 1,200 km. to allow the exchange of energy between member countries. By 1971 there were twenty lines of 110 kV and over, linking the Comecon countries and also Yugoslavia. In 1962 the Comecon countries agreed to establish a central control point in Prague to coordinate the various national control stations. In 1970 the interflow of electric energy amounted to about 13 million MWh, the main suppliers being the USSR and Rumania. The international transmission system was given the name "Mir" (Peace).

In 1960 the Comecon nations produced 17.6 percent of the world's electricity, and by 1970 this had increased to 20.4 percent, or 988 million MWh. By 1975 this is expected to rise to 1,425 million MWh. The contribution of each member state is shown in Table 7.6 in order of importance.

The reliance on thermal power stations will continue, with increasing installation of standard high-capacity units ranging from 200 MW to some of 800 MW in the USSR.[16]

Predictions for future output are notoriously unreliable, both in the USSR and elsewhere. In 1961 Pravda published figures for electricity production in 1970 of 900 to 1,000 million MWh and for 1980 of 2,700 to 3,000 million MWh. Actual output in 1970 was only 740 million MWh, and by April 1972 the planned output for 1980 was a more modest 1,800 million MWh.

Yet it would seem probable that plans as of 1972 will be attained. In 1971 the planned 790 million MWh was surpassed by ten million MWh, and in 1972 the planned 850 million MWh was surpassed by eight million MWh.

Increased capacity from 1970 to 1975 will be predominantly from new thermal stations (47,000 MW out of 67,000 MW). These new stations are mainly equipped with sets of 300 MW capacity, but the Troitsk and Reftinsk stations will have sets of 500 MW, while those of the Zaporozhsk and the Uglegorsk stations will be of 800 MW capacity. The Kostroma station will have a giant 1,200 MW set installed.

By 1975 the construction of the Toktogul, Riga, Kapchagai, and Kanevo hydroelectric power stations was expected to be completed, and the first units of the Ust Ilimsk, Ingursk, Nurek, and Zeya hydroelectric stations were to be in operation.

In 1970 the average amount of conventional fuel used to produce one kilowatt hour was 366 grams; by 1972 this had been reduced to 354 grams, and by 1974 to 348 grams. It is intended to reduce this

MAP 7.1

Electricity Output, by Grid

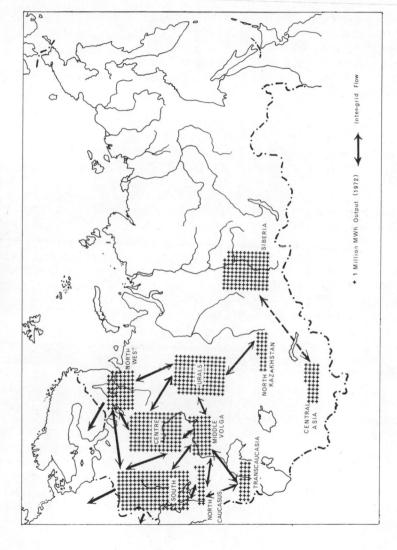

Source: Text, Table 7.5.

TABLE 7.6

Production of Electricity in Comecon Countries

Country	1970			1972	1975
	Output of Electricity (in millions of MWh)	Electricity Produced in Thermal Stations (in percentage)	Utilization of Hydro Resources (in percentage)	Output of Electricity (in millions of MWh)	Planned Output of Electricity (in millions of MWh)
USSR	740.4	83.2	11.4	858.0	1,065.0
Poland	64.5	97.1	28.4	76.4	96.0
East Germany	67.7	97.9	80.0	72.8	89.0
Czechoslovakia	45.2	91.8	40.0	51.4	62.5
Rumania	35.1	90.7	11.7	43.4	59.4
Bulgaria	19.5	88.4	21.5	22.3	30.0
Hungary	14.5	99.4	2.1	16.3	21.5
Mongolia	0.5	NA	NA	NA	0.8

NA: Not available.

Sources: Ekonomicheskaya gazeta no. 6, February 1972: 20; Ekonomicheskaya gazeta no. 4, January 1973: 2.

228

still further by 1975 to 340 to 342 grams per kWh. In the three years from 1971 to 1973 about ten million tons of fuel were saved.[17]

One may conclude from this examination of the Soviet electric power industry that there can be no question of an absolute energy shortage in the USSR in the foreseeable future. The source of primary energy and its rate of utilization will be decided on economic considerations, and in view of the rich reserves of fossil fuels, thermal power stations will be the major producers of electricity until the end of this century and beyond.

NOTES

1. Neporozhnii, P. S., Elektrifikatsiya SSSR 1917-1967, (Electrification of the USSR, 1917-67), Moscow 1967; Vilenskii, M. A., Po Leninskomu puti sploshnoi elektrifikatsii, (On the Lenin road to complete electrification), Moscow 1969; Razvitie elektroenergetiki SSSR, (Development of electric power in the USSR), Moscow 1965.

2. Ekonomicheskaya gazeta no. 6, February 1972: 3.

3. Pospelov, P. N., et al., Velikaya Otechestvennaya Voina Sovetskogo Soyuza 1941-1945 (The Great Fatherland War of the Soviet Union 1941-1945), Moscow 1967: 181, 279, 434.

4. Narodnoe khozyaistvo SSSR v 1969 godu (National economy of the USSR in 1969), Moscow 1970: 192; SSSR v tsifrakh v 1971 godu (USSR in figures for 1971), Moscow 1972: 60-61; Pravda, 22 December 1971: 1.

5. UN Statistical Yearbook 1970, New York 1971: 376; Statistical Abstract of the United States, 1971, Washington 1971: 497; SSSR v tsifrakh v 1971 godu, Moscow 1972: 57-61.

6. Khrushchev, A. T., Geografiya promyshlennosti SSSR (Geography of Soviet industry), Moscow 1969: 208.

7. Ibid.: 218-20; Melnikov, N.V., Toplivno-energeticheskie resursy SSSR (Fuel and power resources in the USSR), Moscow 1971: 42-45.

8. Pravda, 12 June 1971: 1; Pravda, 26 January 1974: 2; Ekonomicheskaya gazeta no. 25, June 1972: 24

9. Ekonomicheskaya gazeta no. 6, February 1972: 24; Ekonomicheskaya gazeta no. 4, January 1973: 2.

10. Voznesenskii, A. N., et al., Energeticheskie resursy SSSR—Gidroenergeticheskie resursy (Energy resources of the USSR—hydropower resources), Moscow 1967: 183; Melnikov, N. V., Mineralnoe toplivo (Mineral fuel), Moscow 1971: 27-29; Rylskii, V. A., Ekonomika mezhraionnykh elektroenergeticheskikh svyazei v SSSR, Moscow 1972: 121-25.

11. Bernshtein, L. V., "Problema ispolzovaniya prilivnoi energii i opytnaya Kislogubskaya PES" (The problem of utilizing tidal energy and the experimental Kislogub tidal power station), Gidrotekhnicheskoe stroitelstvo no. 1, January 1969: 9; BBC Monitoring Service, USSR Weekly Economic Reports (Electricity), 1972-73.

12. This account of the Soviet nuclear energy program has been gleaned from the following sources: Neporozhnii, P. S., et al., Elektrifikatsiya SSSR 1917-1967, Moscow 1967: 162-69; Andreev, V., Atom sluzhit miru (The atom in the service of peace), Moscow 1968: 19-31; Petrosyants, A. M., "Atomnaya nauka i tekhnika k 50-oi godovshchine Velikoi Oktyabrskoi Revolyutsii" (Nuclear science and technology by the 50th anniversary of the Great October Revolution), Atomnaya energiya 23, no. 5, May 1967: 379-93; "Primenenie vodovodyanikh reaktorov v maloi yadernoi energetike" (Using water-cooled reactors in small nuclear power stations), Atomnaya energiya 26, no. 5, May 1969: 403-10; "The USSR Nuclear Power Program: Report of the U.S. Reactor Delegation," Nuclear News 13, no. 10, October 1970: 18-26; Pravda, 23 November 1967: 2; Pravda, 28 July 1969: 1; Pravda, 7 October 1969: 4; Pravda, 27 June 1971: 1; Pravda, 22 December 1971: 1; Pravda, 17 March 1972: 1; Pravda, 22 March 1972: 4; Pravda, 13 April, 1972: 5; Pravda, 15 August 1972: 5; Pravda, 12 June 1973: 4; Pravda, 19 July 1973: 6; Pravda, 22 September 1973: 3; Pravda, 8 December 1973: 2; Sotsialisticheskaya industriya 18 March 1971: 4; Sotsialisticheskaya industriya 27 March 1971: 4; Sotsialisticheskaya industriya 25 March 1973: 2; Izvestiya, 1 October 1969: 4; Izvestiya, 18 July 1973: 6; Izvestiya, 31 December 1969: 2; Sovetskaya Rossiya, 11 April 1973: 1; Sovetskaya Rossiya, 18 August 1973: 2; Sovetskaya Rossiya, 29 December 1973: 4; Trud, 3 July 1973: 1; Novoe Vremya, no. 33, September 1973: 24-25; Novoe Vremya, no. 9, March 1974: 14-15; Jane's Fighting Ships, 1973-74, London 1973; BBC Monitoring Service, Eastern Europe Weekly Economic Report, 5 July 1973; Chemical and Engineering News, 13 November 1972: 22; Guardian, 23 February 1974: 1.

13. Konfederatov, I. Ya., Osnovy energetiki (Bases of power engineering), Moscow 1967: 75-78; Aleksenko, G., et al., "Energiyu solntsa i vetra—v upryazhku" (Using the energy of sun and wind), Pravda, 11 May 1971: 3; Mangushev, K., Prikhodko, N., "Teplo nedr—v delo" (Putting the earth's heat to use), Pravda 30 March 1972: 3; Pravda, 27 April 1973: 3; Pravda, 8 December 1973: 3; Pravda, 15 January 1974: 3.

14. Narodnoe khozyaistvo SSSR v 1969 godu, Moscow 1970: 193; SSSR i soyuznye respubliki v 1970 godu, Moscow 1971.

15. Ekonomicheskaya gazeta no. 6, February 1972; 13; Pravda, 5 November 1973: 2; Sotsialisticheskaya industriya, 22 December 1973: 2.

16. Neporozhnii, P. S., "Sotrudnichestvo energetikov" (Cooperation among energy specialists), Pravda, 18 April 1972: 4; Savenko, Yu., "Razvitie elektroenergetiki" (Development of power engineering), Ekonomicheskaya gazeta no. 6, February 1972: 20; Izvestiya, 1 June 1973: 1.

17. Pravda, 19 October 1961: 3; Neporozhnii, P. S., "Shagi energetiki" (Steps forward in power engineering), Pravda, 22 December 1971: 1; Isaev, V. Ya., "U karty novostroek" (On the map of new construction), Pravda, 2 January 1972: 1; Ekonomicheskaya gazeta no. 4, January 1973: 1; Novoe Vremya, no. 9, March 1974: 15.

A recent Soviet account states, in somewhat general terms, that of the total fuel and energy output in the USSR about half is used for "industrial-technological needs," a third is used for the production of electricity or thermal energy, and more than a tenth is exported.[1] This breakdown is confirmed by the data in Table 8.1. The proportion consumed in the generating of electricity and the proportion of exports have both increased considerably since 1940.

It is difficult to be much more specific than these broad divisions, since in breaking down fuel and energy utilization into separate consumers Soviet sources tend to use many different categories without attempting to define their terminology clearly. Further complications arise because the output of each energy resource and its proportional distribution among consumers changes significantly from year to year. Nonetheless, some careful calculations have been made to determine the major areas of energy consumption, and a fairly comprehensive example is given in Table 8.2, which shows Soviet energy utilization in the 1960s.

Although fluctuations in the demand for energy by consumers are to be expected both in the source of energy and in the quantity required, by considering all sources in conventional fuel tons it is possible to find changes in the structure of total energy utilization from year to year. This has been done in Table 8.3, which shows that despite a considerable increase in the total output of fuels, the proportion consumed by electric power stations grew by only 2.5 percent in the five years from 1965 to 1970.

As yet the share of fuels consumed as a raw material in the chemical industry remains fairly small, although it is rising slowly. Direct fuel consumption by industrial furnaces and by technological or thermal installations has been falling in comparison with the consumption of other major uses.

TABLE 8.1

Changes in the Soviet Energy Balance, 1913-70
(in millions of tons of conventional fuel)

	1913	1940	1960	1970
Total available resources	64.4	283.6	836.5	1,413.4
Output of fuel	48.2	237.7	692.8	1,248.6
Production of hydroelectricity	0.0	0.6	6.3	15.3
Importation	8.0	3.1	10.7	14.1
Other sources	2.4	10.2	32.7	35.4
Remaining from previous year	5.8	32.0	94.0	100.0
Utilization	64.4	283.6	836.5	1,413.4
Consumed	57.6	249.5	678.0	1,130.2
For electric and thermal energy	2.0	44.7	221.2	452.0
For industrial-techno-logical and other requirements (including losses during transportation and storage)	55.6	204.8	456.8	678.2
Exported	1.2	1.1	59.8	163.6
Remaining at end of year	5.6	33.0	98.7	119.6

Source: Sovetskii Soyuz: Obshchii obzor, Moscow 1972: 767.

COMPARATIVE COSTS

A major factor in determining the Soviet energy balance is of course the comparative costs of extracting and transporting the various fuels. Gas and oil are considerably more economical fuels than coal, and their respective shares in the fuel balance have been growing steadily since the 1950s. If the average cost of extracting, processing, and transporting, plus 10 percent of the capital expenditure, for one conventional fuel ton of oil is taken as 100 percent, then the equivalent cost for natural gas is as little as 48 percent and for opencast-mined coal as little as 79 percent, but the cost of coal mined underground is as high as 263 percent.[2]

The comparative cost of extracting, processing, and utilizing one conventional fuel ton of coal, gas, fuel oil, oil shale, and peat in

233

TABLE 8.2

Utilization of Energy Resources
(in percentage)

Use	Coal	Natural Gas	Fuel Oil	Diesel and Motor Fuel	Electric Power
Raw material for processing into other fuels	18.0	3.6	0.4	0.2	—
Generating electricity	26.0	22.1	10.0	14.2	6.4
Thermal energy	17.0	26.1	27.8	1.0	2.7
Mechanical energy	15.2	2.3	28.0	78.4	39.6
Industrial furnaces and other technological installations	8.5	26.6	31.8	3.8	30.7
Housing and municipal services	10.1	13.9	0.6	1.1	12.0
Other uses	6.2	5.4	1.4	1.3	8.6

Source: Lvov, M. S., Resursy prirodnogo gaza SSSR, Moscow 1969: 13.

TABLE 8.3

Major Consumers of Energy Resources, 1965 and 1970
(in percentage)

Consumer	1965	1970
Industrial furnaces and technological installations	18.0	15.0
Thermal installations for industrial purposes	16.0	13.0
Boilers	8.5	9.0
Industrial	4.5	5.4
Heating	4.0	3.6
Power Stations	26.5	29.0
Thermal	23.0	25.4
Hydro and nuclear energy	3.5	3.6
Mobile and on-site installations	15.0	14.3
Consumed as raw materials	3.5	4.6
Lost in extraction, processing, and transportation	1.7	2.2
Other consumers	10.5	10.9

Source: Melnikov, N. V., Mineralnoe toplivo, Moscow 1971: 192.

the USSR in 1970 is shown in Table 8.4. Natural gas is actually cheaper than the local fuels, peat and oil shale, which have almost no transport costs.

The cost of each fuel at source, however, can vary considerably according to the area and conditions of its production, as may be seen in Appendix B. The cheapest sources of gas, oil, and coal are seen to be situated in the eastern parts of the USSR; this necessitates the transportation of these fuels over considerable distances. Comparative transport costs are given in Appendix C, and comparative fuel costs in major areas of consumption are shown in Appendix D.

Taking both cost at source and transport costs into consideration, it has been calculated that the saving per conventional fuel ton that can be made in the European part of the USSR by replacing Donbass or Kuzbass coal (mined underground) with Tyumen natural gas would be 4.5 to 5 rubles; with Central Asian gas, 5 to 5.5 rubles; with fuel oils from refineries west of the Urals, 4 to 7.5 rubles; with fuel oils from refineries east of the Urals, 1 to 2.2 rubles; with electric power from the Itat complex, 2 to 2.8 rubles; with electric power from the Ekibastuz complex, 3 to 4 rubles; with semi-coke from Kansko-Achinsk, 3 to 3.5 rubles; with coal from Kuzbass opencasts, 1 to 1.5 rubles; and with coal from Ekibastuz, 3 to 4.5 rubles.[3]

TABLE 8.4

Estimated Costs of Extracting, Processing, and
Utilizing Major Fuels in 1970
(average for USSR) *

Fuel	Extraction, Processing and Utilization Costs		Transport Costs		Total	
	Rub./Conv. Fuel Ton	As %	Rub./Conv. Fuel Ton	As %	Rub./Conv. Fuel Ton	As %
Coal	9.22	100.0	2.18	100.0	11.40	100.0
Gas	3.12	34.1	1.38	63.5	4.50	39.5
Fuel oil	5.55	60.5	1.55	71.5	7.10	62.1
Oil shale	6.44	70.0	—	—	6.44	56.4
Peat	9.76*	103.0	—	—	9.76*	85.0

*In Rubles per Conventional Fuel Ton
**Includes cost of narrow-guage rail transport from peat bog.
Source: Melnikov, N. V. , Mineralnoe toplivo, Moscow 1971: 178.

Natural gas and, to a lesser extent, fuel oil have such economic advantages over coal that even when the cost of all three is exactly the same at the point of consumption, gas and fuel oil are still considerably cheaper to use. The efficiency of the installation is usually higher, the capital investment lower, and the cost of repairs and maintenance much less. Some examples of the savings that are possible are given in Table 8.5.

It therefore follows from these cost comparisons that there should be considerable expansion of natural gas production in Western Siberia and Central Asia, of oil production in Western Siberia and Western Kazakhstan, and of opencast coal production in the Kuzbass, Kansko-Achinsk basin, and Ekibastuz basin. There must also be increased construction of oil and gas pipelines and of high-voltage transmission lines from large power stations burning cheap coal, which will bring energy from eastern areas to the Urals and the European parts of the USSR. Natural gas is expected to be the major factor in this increased flow of energy, as may be seen in Table 8.6.

ROLE OF NATURAL GAS IN THE ECONOMY

There are two main reasons for examining the role of natural gas in the energy economy of the USSR in more detail than for the other fuels. First, it is the most rapidly developing sector and is expected to be the major fuel in the energy balance by the end of this century. Second, as the most efficient and economical source of energy, it is being adopted in preference to other fuels by a wide range of industrial and domestic consumers.

The structure of natural gas utilization is shown in Table 8.7. Until the demand for gas can be fully satisfied in every region of the USSR, it must be determined which consumers should be given priority. The comparative advantage of gas for each sector of the economy is decided by computing, first, the direct saving made by replacing the former fuel with gas, and second, the less direct advantages of increased productivity from labor and equipment, lower construction costs, improved quality in production, and other such factors.

In an industry such as glass making, which requires a fuel in the form of gas, it will be of obvious advantage to dispense with equipment for manufacturing gas from other fuels. In many high-temperature engineering processes the direct use of natural gas can replace electricity with considerable improvement in total energy efficiency.

In industries such as engineering, metalworking, oil refining, and brick making, natural gas allows the introduction of newer, more efficient techniques.

TABLE 8.5

Economic Effect of Replacing Coal with Natural Gas or Fuel Oil
(assuming that cost to consumer per conventional fuel ton of coal, gas, and oil is the same)

Installation	Savings per Conventional Fuel Ton (in rubles)	
	Natural Gas	Fuel Oil
Industrial Furnace	8.0 to 15.0	5.5 to 8.0
Heating Boilers	6.0 to 10.0	4.5 to 7.0
Industrial Boilers	3.2 to 4.5	2.1 to 2.8
Boilers of Large Power Stations	2.0 to 2.3	1.6 to 1.8
Replacing Coal-Fired Steam Locomotives with Diesel Locomotives	—	15.0 to 18.0

Source: Adapted from Ekonomicheskaya entsiklopediya: Promyshlennost i Stroitelstvo, Moscow 1965, vol. 3: 418.

TABLE 8.6

Predicted Energy Flow from the East to the West of USSR
(in millions of conventional fuel tons)

Energy Source	1965	1970 (estimated)	Future
Natural gas	12	48	285 to 325
Oil	—	—	120
Energy coal	38	39 ⎫	100
Concentrate for coking	20	26 ⎭	
Electric Power (in millions of MWh)	—	—	200
Total	70	113	565 to 700

Source: Melnikov, N. V., Mineralnoe toplivo, Moscow 1971: 194.

237

TABLE 8.7

Structure of Natural Gas Utilization in 1965 and Estimated for 1972

Utilization	1965 (in thousands of millions of cu.m.)	1965 (in percentage)	1972 (estimated) (in thousands of millions of cu.m.)	1972 (estimated) (in percentage)
Municipal and household	14.9	11.5	25.5	11.4
As raw material	8.3	6.5	18.3	8.2
Chemical processing	3.7	2.9	10.2	4.6
Production of liquid gas and petrol	2.0	1.6	4.8	2.1
Production of carbon black	2.6	2.0	3.3	1.5
Technological	41.9	32.5	69.2	31.3
Iron and steel processing	13.0	10.0	27.5	12.4
Nonferrous metals processing	1.0	0.7	2.0	0.9
Manufacture of industrial building materials	11.0	8.6	16.7	7.6
Engineering and metalworking	4.6	3.6	8.8	4.0
Oil refining	3.6	2.8	4.7	2.1
Others	8.7	6.7	9.5	4.3
In energy production	59.0	45.7	92.8	42.1
For Gas industry's own needs and for export	5.2	3.8	15.5	7.0
Total	129.3	100.0	221.0	100.0

Source: Adapted from Energeticheskie resursy SSSR: toplivno-energeticheskie resursy, Moscow 1968: 543.

In a third group of industries, such as cement manufacturing, there is no immediate improvement in the techniques used; but furnace productivity increases, fuel consumption drops, and the quality of production rises.

In electric power production, natural gas improves boiler efficiency and reduces fuel costs.[4]

The advantages of using natural gas in certain industries are discussed in more detail below.

In the chemical industry natural gas is the most efficient raw material for producing ammonia, ethylene, acetylene, and methanol. It is replacing alimentary raw materials by providing ethyl alcohol for synthetic rubber; this saves millions of tons of grain, potatoes, and other foods. Using natural gas reduces the cost of solvents, resin, formalin, phenol, acetone, and other valuable chemicals. Natural and oil-well gas are being used in the production of nitrous fertilizers, in factories constructed in areas of high demand.

Large-scale use of natural gas as a raw material began in 1958 at the Novomoskovsk and Lisichansk plants and later was adopted at the Chirchik, Rustavi, Shchekino, Nevinnomyssk, Salavat, and Fergana plants. By 1965 over 40 percent of methyl alcohol and 60 percent of all ammonia produced in the USSR was from natural gas, at a cost 35 to 50 percent less than that of using solid fuels. Where the production of a ton of ammonia consumes 2 tons of fuel oil or 3.4 tons of coal in conventional fuel, only 1.6 tons of natural gas is required.

In the iron and steel industry natural gas can to some extent replace expensive coke and low-sulphur fuel oil. Yearly consumption of natural and oil-well gas in Soviet iron and steel production increased from 1,693 million cu.m. to 17,637 million cu.m. in the period from 1958 to 1965 and has continued to rise sharply every year, reaching 26,400 million cu.m. in 1969. By 1965 it provided 46.7 percent of the fuel mix in steel smelting and 31 percent of the fuel mix in producing rolled steel, although still only 9.6 percent of the fuel mix in cast iron production. With the conversion of the Ural works to natural gas the proportion has grown considerably. In 1969 as much as 85 percent of cast iron and 80 percent of steel was produced using natural gas in some stage of the process. The average saving for every 1,000 cu.m. of natural gas used to replace other fuels in the iron and steel industry has been from 17 to 23 rubles, according to the region.

In the nonferrous metals industry natural gas is used in more than thirty different processes, reducing fuel consumption and increasing technological efficiency. It is of particular importance in the production of copper, lead, zinc, and tin as both a source of energy and as part of the manufacturing process; in lead and zinc plants it has replaced coal fires, with a 20 percent saving in fuel consumption. The saving on every 1,000 cu.m. of natural gas used in preference to solid fuels in the aluminum industry, for instance, is 3.2 to 8.7 rubles.

In the cement industry natural gas has been the major fuel since 1963. Cement factories consume 70 percent of the natural gas supplied to construction enterprises. Gas cuts fuel expenditure by 7 to 10 percent. In 1958 the components of the fuel mix in the cement industry were coal 62 percent, natural gas 29 percent, fuel oil 7 percent, and oil shale 2 percent; by 1965 coal had dropped to only 24 percent, while fuel oil had risen to 16 percent and gas to 58 percent. Among the other advantages of natural gas are increased furnace productivity (7 to 12 percent); higher labor productivity (5 to 10 percent); improved quality because of the absence of ash deposited by solid fuel; and reduced capital investment because there is no longer a need to transport, store, and prepare coal (5 to 10 percent). The saving on every 1,000 cu.m. of natural gas used in preference to solid fuels is 13 to 17 rubles. In 1970 industries making building materials consumed 19,200 million cu.m. of natural gas.

In the glass industry fuel is consumed in the form of gas; by using natural gas the expense of solid fuel gasification can be avoided. One ton of glass requires (in conventional fuel) 460 to 640 kg. for natural gas; 620 to 690 kg. for fuel oil; or 790 to 1,850 kg. for solid fuel. Changing factories from solid fuel to natural gas reduces fuel consumption by over 25 percent and in addition saves 18 rubles for every 1,000 cu.m. consumed, because natural gas costs less than other fuels.

In the ceramics industry using natural gas reduces fuel consumption by 5 to 8 percent compared with fuel oil and by 20 to 25 percent compared with generator gas; there are also considerable savings in capital investment and wages because the production process is greatly simplified.

In engineering and metalworking partial replacement of coke by natural gas (up to 50 percent) has improved the productivity of cupola furnaces by 17 to 20 percent and brought savings of 50 to 60 rubles for every 1,000 cu.m. consumed. In many technological processes natural gas furnaces can replace electric furnaces, with considerable savings, by eliminating the intermediate stage of generating electricity. In 1970 this sector of industry consumed about 20,500 million cu.m. of natural gas, over 10 percent of the total Soviet output.

For light industries and food industries an important factor is the improvement in quality that can be achieved by using natural gas burners in preference to steam heat in, for example, baking, or in the thermal processing of textiles. Fuel consumption can be reduced and the productivity of labor and equipment increased.

In energy production there are also many advantages to be gained by using natural gas. The efficiency of industrial boilers can be increased by 5 to 25 percent, according to the type of boiler and the type of fuel being replaced. Steam production can be significantly raised at peak periods, which may make it unnecessary to construct auxiliary boilers. Personnel may be reduced by as much as 50 percent. Gas-fired boilers are much cheaper to install than solid-fuel boilers. Automation of control and servicing is simpler with gas than with other fuels. The saving from using gas rather than coal is about 11 to 17 rubles for every 1,000 cu.m.

The proportion of the Soviet fuel output that is consumed in power stations has been steadily increasing, from 25.6 percent in 1962 to over 30 percent at the beginning of the 1970s. Hard coal is still the major component in the fuel mix of power stations, but the share of natural gas has been growing rapidly; in 1958 only 9,600 million cu.m. was supplied to power stations, but in 1965 it was as much as 35,700 million cu.m. The use of gas to generate electricity can be made much more economical by the construction of gas-fired power stations that can use fuel oil or coal at peak periods when gas supplies are inadequate. Existing power plants that now use solid fuels can be converted to the same system of operating. The cost of constructing such gas-fired power plants is approximately 15 percent less than that of constructing power plants fired only by solid fuels. The personnel required to run a gas-burning station with a fuel-oil reserve is 33 to 35 percent less than for an anthracite-based plant and 40 to 42 percent less than for one based on brown coal. Electric power consumed for the stations' own needs drops from 6 to 8 percent for solid-fuel plants to 4 percent for gas-burning plants. The saving on every 1,000 cu.m. consumed in the new gas-and-oil-fired stations, compared to coal-burning plants, is calculated at 11 to 12 rubles, while in stations converted from coal to natural gas it is put at 9.5 to 10.5 rubles. The advantages of using gas in heating and power plants are similar, and the saving is often even greater than for other thermal power stations.

Domestic consumption of natural gas has been growing steadily with the expansion of the pipeline network. The number of households that are benefiting from this clean and convenient fuel rises every year. (See Table 8.8.) By 1970 there were 1,720 large towns, 1,866 small towns (poselok gorodskogo tipa), and 38,070 villages being supplied with gas through the pipeline network, while some country areas were supplied with liquified gas.

Freedom from pollution is only one of the advantages of changing to natural gas from the solid fuels. The efficiency of an average domestic gas cooker is 65 to 70 percent, compared to 30 to 35 percent for solid fuel. The saving on fuel cost for an average family can be 30 to 40 rubles a year; the savings of time and effort, although incalculable, are very great. The savings realized by supplying towns with piped gas are increased by expanding consumption to the maximum. In addition to domestic cooking, gas should be used for water and space heating, both in homes and in communal buildings such as restaurants, hotels, places of work, and places of entertainment; it should, of course, also be adopted in the town's industries.

TABLE 8.8

Number of Homes Supplied with Gas, 1970 and 1975

Region	Number of Flats (in thousands at end of year)	
	1970	1975 (estimated)
USSR	23,366	41,496
RSFSR	11,671	22,271
Ukrainian SSR	4,883	8,183
Belorussian SSR	808	1,508
Lithuanian SSR	375	657
Latvian SSR	369	575
Estonian SSR	223	358
Moldavian SSR	352	603
Georgian SSR	594	986
Azerbaidzhan SSR	521	892
Armenian SSR	319	409
Uzbek SSR	1,059	1,932
Kirgiz SSR	311	561
Tadzhik SSR	190	431
Turkmen SSR	247	409
Kazakh SSR	1,456	2,488

Source: Calculated from data in Gazovaya promyshlennost no. 12, December 1972: 6-7.

The cost per household of installing a piped gas system for domestic use is greatly reduced where buildings have four or more floors, in densely populated housing estates. Since such accommodations are also much cheaper to build than small, detached homes it is scarcely surprising that large-scale construction of multistory blocks of flats has been adopted in an attempt to remedy the acute shortage of housing. The saving on every 1,000 cu.m. consumed in the communal-domestic sector varies from 6 to 14 rubles.[5]

Natural gas is even being used as a fuel to provide motive power. It was reported in March 1973 that there were over 1,500 trucks in Moscow that consumed liquid gas instead of gasoline in an attempt to reduce air pollution. Liquid gas was used as a transport fuel during the war, when oil products were needed for tanks and other military vehicles, but it has now proved to be more economical than it was then; each gas-powered vehicle saves some 500 rubles a year on fuel alone, and there are also considerable savings because of longer engine life.

In addition to the two models of trucks, the ZIL-166 and GAZ-51ZH, that have been adapted to liquid gas, more specially designed trucks, buses, and taxis are planned that will use this cleaner fuel. The first thousand of these vehicles are to be on the Moscow streets in 1974. A major problem has been the lack of service stations that can supply liquid gas. In 1973 there were only eight such stations, and four of them had been closed down. It was planned to import some 95 liquid gas pumps to rectify this situation, and experts hoped to extend this form of transport to the other major Soviet cities. In the 1974-76 period it is expected to have 35,000 gas-powered vehicles in Moscow alone, but still very few service stations; under fifty by 1976.[6]

CONSERVATION OF ENERGY

The rational utilization of fuels is closely linked to problems of conserving energy; it is estimated that at present over half the world's energy production is wasted through inefficient methods of consumption. Soviet planners are becoming increasingly aware of these losses to the economy, and industries are now often expected to economize on energy consumption as part of their plan.

In the transportation and unloading of coal and peat a loss norm of .8 percent has been set, but actual losses have been closer to 3 percent. By cutting losses to the established norm in the 1975-80 period it is hoped that about 80 million tons of coal will be saved. Similarly the loss of coal through poor storage at the minehead and near large consumers has been about 7 percent, rather than the planned norm of 3.4 percent. When coal is being transported by rail as much as 1.2 to 1.5 tons per car (2 to 4 percent of the total) can be lost, a third through cracks and the remainder blown away. In 1965 the cost of cleaning the railroad tracks was about 50 million rubles. This waste could be eliminated by the construction of specially designed rolling stock similar to that used in Western countries and by expanding the building of coal and peat stores to protect the fuel from the weather.

There is also considerable waste in the utilization of incidental oil-well gas. Of the 120,000 million cu.m. produced between 1870 and 1970, it is calculated that only about 53 percent has been put to any use, the remainder being vented and flared. In the 1971-75 period it is expected that 138,000 million cu.m. will be produced, of which 83 percent will be utilized. In the United States, by contrast, less than 10 percent of oil-well gas is wasted. This situation is gradually improving in the USSR with the increased building of processing plants and power stations based on incidental gas.

At the beginning of the 1970s the efficiency rate for the consumption of fuels in Soviet industry was as low as 32 to 33 percent. This can be greatly improved by raising the utilization of secondary energy resources, such as gas from industrial furnaces and blast furnaces, and heat from cooling furnaces and from the dry quenching of coke.

The wider use of gas and oil and the adoption of more efficient thermal equipment cut expenditure on fuel in the 1960-65 period by 11 percent. Of the total saving of 58.1 million tons of conventional fuel, 40 million tons were saved in power stations, 8.7 million tons in industrial furnaces, 4.7 million tons in blast furnaces, and 4.7 million tons in open-hearth furnaces.

In railway transport there has been a steady change from steam to diesel and electric locomotives. In 1958, when some 60 percent of the total railway haulage was by steam locomotive, 73 million tons of conventional fuel were consumed. By 1965, although railway haulage had increased by 46 percent, the amount of fuel consumed had dropped to 41 million tons, including fuel for the production of electricity, because the proportion of steam locomotives had dropped to 15.5 percent. By 1971 very little railway haulage was by steam locomotives; 49.6 percent was by electric locomotive and 47.8 percent by diesel.

The massive construction of private cars now underway in the USSR is certainly less sensible from the point of view of conserving energy. Although passenger discomfort from crowding in buses, trolley-buses, streetcars, and subways has been considerable, the solution is surely to increase the number of these vehicles in service, rather than risk the fate of Western cities by crowding the roads, which are already inadequate, with private cars.

The changing fuel balance for boilers and furnaces is shown in Table 8.9. The expanding share of natural gas and, to a lesser extent, of fuel oil, with the corresponding drop in importance of the less economical fuels, will be noticed.

Power stations require over 30 percent of the total boiler fuel consumed, and this proportion is expected to grow to 45 percent by 1990. The amount of energy they expend on their own requirements (6 percent on average) is considered too high. Regional and industrial boilers consume over 10 percent of the total fuel resources, often at a low (50 to 74 percent) efficiency rate. The planned efficiency of 300 MW turbines, for example, was to be 70 to 75 percent, but has proved to be only 50 percent. In 1968, because of poor-quality equipment and faulty installation, such turbines produced some 20 to 25 MWh less than planned. Efficiency will be raised by expanding the number of large, improved heat and power plants.

Ferrous and nonferrous metals processing consumes over 110 million tons of conventional fuel a year. Improved technology and equipment has reduced the fuel consumption per ton produced in the

TABLE 8.9

Fuel Structure for Boilers and Furnaces, 1960-70
(in percentage)

Fuel	1960	1965	1970 (estimated)
Coal	54.6	45.0	38.4
Coke and byproducts	9.6	8.4	8.0
Fuel oil	11.6	13.3	14.9
Natural, oil-well, and liquid gas	9.4	22.1	28.5
Manufactured gas	6.6	5.1	6.1
Peat	3.8	2.3	2.1
Oil shale	0.4	0.6	0.7
Wood	4.0	3.2	2.6

Source: Melnikov, N. V., Mineralnoe toplivo, Moscow 1971: 199.

period from 1960 to 1965: for steel, fuel consumption is reduced from 186.3 to 172 kg.; for rolled steel, from 153.5 to 147.1 kg.; and for smelting cast iron, from 763 to 633 kg. The ratio is still being improved, especially through the increased use of natural gas.

In the cement industry the replacing of furnaces 60 to 75 m. in length by new ones 170 to 185 m. in length has been proven to reduce fuel consumption by 30 to 40 percent.

In house building greater care is being taken with thermal insulation, since the additional cost of insulating a high block of flats properly can be recovered in a few years through the saving of fuel.

Considerable effort is being directed at reducing the amount of oil and gas lost during extraction, transportation, and refining. In the first place, only 40 to 60 percent of the oil is at present extracted from the deposit; second, several million tons a year are lost in transportation; third, refineries use 14 to 15 percent of gas or fuel oil for their own requirements; and fourth, the efficiency rate of furnaces is only 55 to 65 percent. Indeed, the total losses of oil and gas from source to refinery have been estimated at 15 to 20 percent. In 1969, for example, over 40 million tons out of the total output of 328 million tons were consumed by the oil industry for its own needs.

Great attention is therefore being paid to energy conservation in the USSR; the present low efficiency in energy utilization would be unthinkable by the end of the century, when fuel consumption may reach 6,000 million tons of conventional fuel, and electric power production 8,000 million MWh a year.[7]

REGIONAL NEEDS

One of the major problems facing Soviet planners is the fact that industry and population are concentrated in the European USSR and the Urals, which have only 12 percent of the potential energy resources and 25 percent of the proven reserves. This share includes 18 percent of the potential coal resources, 32 percent of the natural gas, and 17 percent of the hydropower. In 1965 the Urals and the area to the west could satisfy 91 percent of their own energy requirements; by 1970 this dropped to 87 percent; and by the 1980s it will probably be as low as 70 percent, requiring the importation of an additional 300 to 400 million tons of conventional fuel from other areas.

There are various steps that can be taken to deal with this threatening deficit. In the Urals and European USSR the growth of industries that are large consumers of energy must be strictly limited. Electric power should be produced in the future in nuclear stations, large-scale and more efficient thermal stations, or hydroelectric stations, which can supply additional power at peak times. The inter-regional electricity grid should be extended, to ensure uninterrupted supplies to consumers.

Oil and gas pipelines and high voltage transmission lines must be further developed, to bring energy from eastern areas. (See Table 8.6.) Energy-intensive industries will have to be established east of the Urals, even though this may require transportation of raw materials from western areas. It has been calculated that when such industries have been established in Eastern Siberia the savings have been considerable, even allowing for the cost of transporting the raw materials to the east and the finished product back west. The saving per ton of nickel, for example, is 150 rubles; for synthetic rubber, 80 rubles; for synthetic resins, 30 rubles; and for aluminum, 30 rubles.

The alternative of transporting energy westwards is expensive. Supplying the Center with one conventional fuel ton costs 4 to 6 rubles for Central Asian gas, 5 to 7 rubles for Tyumen gas, 9 to 10 rubles for Kuznetsk coal, and 8 to 9 rubles for Ekibastuz coal. Transportation tends to double or triple the cost of the energy at its source.

These factors have led to the planning of new industrial complexes in the east of the USSR, especially in central and western Siberia, northern Kazakhstan, southern Tadzhikistan, and the Far East.

The steady rise in the importance of the eastern areas of the USSR in the production of energy (Table 8.10) is therefore being matched by an expansion of the energy-intensive sectors of industry east of the Urals.[8]

Any discussion of economic regions in the USSR is complicated by the fact that there have been several reorganizations of boundaries in recent years, and it is possible that the eighteen divisions at the

TABLE 8.10

Growth in Energy Production in
Eastern Areas, 1940-71
(in percentage of total USSR output)

Type of Fuel	1940	1960	1965	1970	1971
Electricity					
Including Urals*	21.7	38.4	39.3	37.9	37.9
Excluding Urals†	9.2	21.6	24.3	26.2	26.4
Oil					
Including Urals*	6.9	9.6	12.1	24.9	28.4
Excluding Urals†	6.3	7.2	7.1	18.1	21.5
Gas					
Including Urals*	0.6	3.5	15.4	30.9	33.2
Excluding Urals†	0.5	2.4	14.5	29.8	31.3
Coal					
Including Urals*	35.9	46.7	48.8	50.6	51.2
Excluding Urals†	28.7	35.9	39.3	43.2	44.3

*All eastern parts of USSR including the Urals.
†Eastern parts of USSR excluding the Urals.

Source: Narodnoe khozyaistvo SSSR, 1922-1972, Moscow 1972: 142.

beginning of the 1970s will be changed to seven large units. There would then be three in the European part of the USSR the North-Central, the Southern, and the Volga-Ural regions; and four in the eastern areas, Central Asia, Kazakhstan, Siberia, and the Far East.[9]

The major division will doubtless continue to be between the underdeveloped east with huge natural resources but sparse population, and the European part of the USSR, with a relatively large population but with mineral resources approaching exhaustion. Table 8.11, for example, shows that in 1965 the areas east of the Urals had 90 percent of the Soviet energy resources, yet produced under 23 percent of the total output of fuel. By way of contrast, slightly more was extracted in the Ukraine, which had under 3 percent of reserves, while the Volga and Ural regions, with under 2 percent, accounted for some 33 percent of the fuel produced in that year.

The regions with the greatest deficit in energy were the Center, the Urals, and the Northwest, but in fact all the European regions, except for Volga, North Caucasus, and Transcaucasia, also consumed more energy than they could produce. (See Map 8.1.)

The fuel balance of the regions as of 1965 is shown in Table 8.12. Coal was still the major source of energy in the Northwest, Center, Central Black Earth, Ural, Siberian, Far East, and Kazakhstan regions. In most of these regions oil was of increasing importance; it was already the main source of energy in the Volga, Volgo-Vyatka, Baltic, Belorussian, Transcaucasian, and Central Asian regions. Natural gas provided over 20 percent of the fuel mix in the Center, Volga, Ukraine, Transcaucasia, Central Asia, and North Caucasus regions, and it was the major fuel in most of the latter region. Peat was an important factor in the Center and Belorussia, oil shale in the Baltic, and wood in the Northwest, Volgo-Vyatka, and East Siberian regions.

Since 1965 the share of oil and especially of natural gas has tended to increase in most regions. By 1967 natural gas was providing 45 percent of fuel requirements in the Center, 47 percent in Transcaucasia, over 50 percent in the North Caucasus, and over 30 percent in Central Asia and Kazakhstan. By 1970 its share in the fuel balance of the last two regions had climbed to over 60 percent. As the extraction of Tyumen gas increases, it will provide over 45 percent of the fuel requirements of the Ural, Volgo-Vyatka, Northwestern, Baltic, Belorussian, and Central regions.[10] This is demonstrated by Table 8.13.

There can be no single solution, however, to the threatening energy deficit in the western areas. To supply the necessary energy from Tyumen gas alone would require the construction of 14 pipelines of 1,400 mm., or six of 2,500 mm., over a distance of almost 4,000 km. in order to deliver about 350,000 million cu.m. every year. The construction of railways to deliver 400 million conventional fuel tons of coal from the Kuzbass or Kansko-Achinsk basin is also clearly not practicable. Even less feasible are other alternatives, such as building nuclear power stations with a total capacity of 200 million kW or planning thirty high tension transmission lines of 1,500 kV over 3,000 km. in length.[11]

Soviet planners are therefore attempting to find the best possible combination of imported energy and local resources for each region. The optimum solution tends to change from year to year in accordance with resource availability, transportation facilities, and comparative costs. Decisions in this sphere are also very strongly influenced by export policies.

TABLE 8.11

Production and Consumption of
Energy Resources, by Region, 1965

Region	Total Energy Re- sources*	Hydro- power*	Percent- age of USSR Reserves	Percent- age of USSR Output of Fuel	Percent- age of USSR Con- sumption of Fuel
Northwest	263	3.5	4.2	2.3	6.1
Center	13	0.5	0.2	2.2	10.5
Volgo-Vyatka	3	0.5	—	0.2	2.7
Central Black Earth	—	0.1	—	—	2.9
Volga	65	2.2	1.1	27.0	8.4
North Caucasus	66	3.8	1.1	10.9	4.3
Ural	32	1.5	0.5	5.9	12.7
West Siberia	859	7.4	13.9	9.3	7.0
East Siberia	2,627	29.9	42.6	3.3	5.3
Far East	1,898	34.2	30.3	2.1	3.4
Ukraine	146	1.3	2.4	23.4	20.7
Baltic	6	0.4	0.1	0.7	2.7
Transcaucasia	24	6.7	0.4	4.1	3.0
Central Asia	71	16.4	1.2	4.4	3.0
Kazakhstan	120	5.3	1.9	3.8	4.4
Belorussia	3	0.2	0.1	0.4	2.3
Moldavia	—	0.1	—	—	0.6
Total for USSR	6,201	114.0	100	100	100
European USSR	588	19.3	9.5	71.2	64.2
Urals	32	1.5	0.5	5.9	12.7
Asian USSR	5,581	93.2	90.0	22.9	23.1

*In thousands of millions of tons of conventional fuel.

Source: Compiled from Probst, A. E., et al., Razvitie toplivnoi bazy raionov SSSR, Moscow 1968: 36, 45, 48.

MAP 8.1

Regional Energy Flow

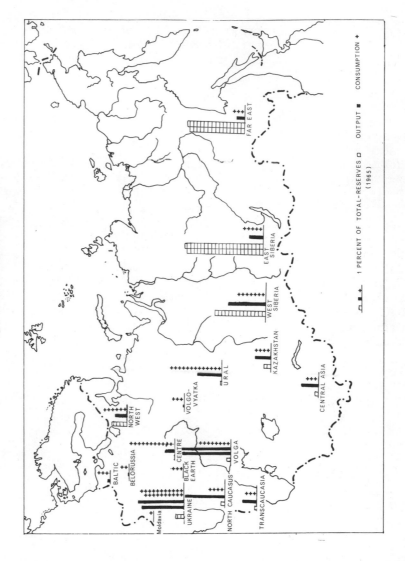

1 PERCENT OF TOTAL—RESERVES ▢ OUTPUT ■ CONSUMPTION +

(1965)

Source: Text, Table 8.11.

TABLE 8.12

Fuel Balance of Main Economic Regions, 1965
(in percentage)

Region	Natural and Incidental Gas	Oil and Oil Products	Coal and Coal Briquettes	Coke and Coke Fines	Peat and Peat Briquettes	Oil Shale	Wood	Secondary Energy Resources
Northwest	14.1	27.4	34.2	3.9	4.1	1.0	12.6	2.7
Center	27.3	24.3	29.3	3.0	17.9	—	6.1	2.1
Central Black Earth	14.3	35.5	35.8	9.3	0.9	—	2.3	1.9
Volgo-Vyatka	18.1	27.9	26.9	0.8	9.1	—	14.3	2.9
Volga	27.3	45.2[a]	20.9	0.8	—	0.3	3.9	1.6
North Caucasus	35.4	31.9	27.9	0.7	—	—	4.1[b]	NA
Ural	13.0	12.0[c]	50.0	NA	NA	NA	NA	NA
West Siberia	—	23.2	60.0	6.8	—	—	5.7	4.3
East Siberia	—	27.5	61.1	0.1	—	—	10.0	1.3
Far East	1.8	29.0	61.1	0.3	—	—	7.6	0.2
Ukraine	23.0	17.1	37.8	15.2	1.0	—	1.5	4.1
Baltic	8.7	36.3	22.2	0.4	6.9	15.6[d]	8.2	1.7
Trans-caucasia	32.5	52.0	9.5	2.7	—	—	0.9	2.4[b]
Central Asia	26.5	42.6	26.5	—	—	—	0.6	3.8[b]
Kazakhstan	1.9	35.8	54.2	4.6	—	—	1.7	1.8
Belorussia	13.4	31.7	25.8	0.7	19.4	—	7.6	1.4

NA: not available:

[a]Includes 5.9 percent consumed as gas from oil refining.
[b]Includes other sources of energy.
[c]Fuel oil only.
[d]In Estonia alone oil shale was about 60 percent of the fuel balance.

Source: Compiled from data in Probst, A. E., et al., Razvitie toplivnoi bazy raionov SSSR, Moscow 1968: 69-316.

251

IMPORTANCE IN TRADE

The importance of energy in Soviet balance of trade figures has been growing steadily in recent years. While the share of fuels and electric power in the total value of exports has increased from 3.8 percent in 1950 to 17.6 percent in 1972, their share in imports has dropped from 11.8 percent in 1950 to a mere 2 percent in 1970, though rising slightly to 3 percent in 1972. The comparative importance of each energy source in the total value of exports is shown in Table 8.14.

In actual terms the increase in the export of fuels, especially oil and oil products, is even more striking. (See Table 8.15.) The proportion of the total Soviet oil output that is exported has been very high indeed, about 30 percent in the early 1970s. As domestic consumption increases because of such developments as the expanded production of private cars, the problem of meeting the oil requirements of Comecon countries and of paying for Western technology will become acute.

Per capita energy production in the USSR is still well below that in the United States. In 1970 the comparative figures were as follows:[12]

	USSR	United States
Electricity (in kWh)	3,052	8,423
Oil (in kg.)	1,452	2,313
Gas (in cu.m.)	822	3,023
Coal (in kg.)	2,379	2,640

TABLE 8.13

Expected Fuel Balance of Selected Regions by 1980,
Showing Influence of Tyumen Gas
(in percentage)

Region	Gas	Fuel Oil	Coal	Other Fuels
Northwest	62	12	8	18
Center	58	13	12	17
Ural	51	6	19	24
Baltic	47	21	5	27
Volgo-Vyatka	45	18	14	22
Belorussia	38	28	14	20

Source: Gazovaya promyshlennost no. 11, November 1969: 13.

TABLE 8.14

Energy Resources as Proportion of Total Value
of Soviet Exports, 1950-72
(in percentage)

Type of Fuel	1940	1950	1960	1970	1971	1972
Oil	—	0.4	5.0	7.2	8.5	8.6
Oil Products	13.2	2.0	6.9	4.3	4.8	4.4
Coal (hard and anthracite)	0.1	0.7	3.2	2.4	2.6	2.5
Coke	—	0.7	1.2	0.8	0.9	0.9
Gas	—	—	—	0.4	0.5	0.4
Electric Power	—	—	—	0.4	0.6	0.6
Total	13.2	3.8	16.3	15.5	17.9	17.6

Sources: Vneshnyaya torgovlya SSSR, 1918-1966, Moscow 1967: 16-17, 74-75;
Vneshnyaya torgovlya SSSR za 1972 god, Moscow 1973: 20.

TABLE 8.15

Soviet Export of Fuels and Electricity, 1950-72

Type of Fuel	1950	1955	1960	1965	1970	1972
Crude Oil[a]	0.3	2.9	17.8	43.4	66.8	76.2
Oil Products and synthetic liquid fuel[a]	0.8	5.1	15.4	21.0	29.0	30.8
Gasoline[a]	0.3	1.4	2.8	2.4	3.5	4.5
Diesel fuel[a]	0.1	1.3	5.2	7.4	11.4	12.5
Fuel oil[a]	0.1	1.5	5.8	9.7	11.4	11.1
Lubricating oils[a]	0.07	0.15	0.4	0.3	0.3	0.2
Kerosene and jet fuel[a]	0.23	0.64	1.8	1.5	2.1	2.3
Coal (hard)[a]	1.0	2.9	10.5	18.8	20.0	20.2
Anthracite[a]	0.1	1.4	1.8	3.3	4.3	4.2
Coke[a]	0.7	1.6	2.6	3.8	4.2	4.5
Gas[b]	0.08	0.14	0.24	0.39	3.30	5.1
Electricity[c]	0.00	0.00	0.03	1.52	5.35	7.5

[a]In millions of tons.
[b]In thousands of millions of cu.m.
[c]In thousands of millions of kWh.

Sources: Vneshnyaya torgovlya SSSR 1918-1966, Moscow 1967: 80-83. Statisticheskii ezhegodnik stran-chlenov SEV 1971, Moscow 1972: 369-70. Statisticheskii
ezhegochink stran-chlenov SEV 1972, Moscow 1973: 382. Vneshnyaya torgovlya SSSR
za 1972 god, Moscow 1973: 27.

253

The difference in energy consumption per capita is even greater because of the fact that, unlike the United States, the USSR exports more energy than it imports. It can be argued, of course, that a more equitable distribution of the world's wealth is greatly to be desired and that some limit should be placed on the energy consumed by any one nation or even individual. Nonetheless, the fact remains that if the Soviet government wishes to raise the standard of living in the USSR to the level enjoyed in the developed nations of the West, it will have great difficulty in continuing to export such a high proportion of its energy output.

The main importing countries for Soviet oil and oil products are shown in Table 8.16. They are mostly Comecon countries, but Finland, Italy, West Germany, Sweden, France, and Japan also receive large quantities. Supplies to China stopped because of the Sino-Soviet dispute.

Soviet hard coal and anthracite are also exported mainly to the Comecon nations, with Japan, Italy, and France being the major capitalist importers. (See Table 8.17.) In 1972 the 4.5 million tons of coke exported by the Soviet Union went mainly to Romania (1.2 million tons), East Germany (1.3 million tons), Hungary (.6 million tons) and Finland (.6 million tons).[13]

Of the 7,500 million kWh of electric power transmitted by the USSR to other countries in 1972, most went to Hungary (4,113 million kWh) and Czechoslovakia (1,081 million kWh). Finland, Poland, and Bulgaria are also connected to the Soviet grid to enable them to receive additional power at peak periods.[14]

Exporting natural gas was still in its initial stages in 1972 when 5,100 million cu.m. was pumped through the Bratsvo pipeline to Czechoslovakia (1,937 million cu.m.), Austria (1,633 million cu.m.), and Poland (1,500 million cu.m.). In the same year almost double that amount was received by the USSR from Iran (8,197 million cu.m.) and Afghanistan (2,849 million cu.m.). The export of gas to Bulgaria began in 1973, and, as was mentioned in Chapter 2, agreements have been signed with Italy to supply it with 100,000 million cu.m. of Soviet gas over twenty years, beginning in 1972. A similar agreement with West Germany called for the completion of a pipeline from the USSR in 1973; the flow of gas will total 52,000 million cu.m. over the following twenty years. This pipeline will probably be extended to France.[15]

In 1972 Soviet exports of oil and oil products totaled 107 million tons, about half of which went to the Comecon nations. The other major importers were Finland (8.6 million tons), Italy (8.4 million tons, or almost 2 million tons less than in 1970), West Germany (6.2 million tons), Sweden (4.4 million tons), France (3.1 million tons), Belgium (2.5 million tons), Holland (2.4 million tons), and Japan (just over 1 million tons, 2 million tons down from the previous year). Exports of electric power rose from 7,000 million kWh in 1971 to

TABLE 8.16

Main Countries Importing Soviet Oil
and Oil Products
(in millions of tons)

Country*	Crude Oil 1955	Crude Oil 1960	Crude Oil 1965	Oil Products 1955	Oil Products 1960	Oil Products 1965	Crude Oil and Oil Products Combined 1970	Crude Oil and Oil Products Combined 1971	Crude Oil and Oil Products Combined 1972
Czechoslovakia	0.4	2.4	6.0	0.2	0.3	0.4	10.5	11.8	12.9
East Germany	0.7	1.8	4.9	—	0.4	0.5	9.3	10.4	11.5
Poland	0.4	0.7	3.2	0.2	1.4	1.5	8.6	9.5	11.1
Italy	0.1	3.9	6.5	0.1	0.8	0.7	10.2	9.0	8.4
Finland	—	0.8	1.9	0.6	1.4	2.6	7.8	8.5	8.6
Bulgaria	—	—	2.2	0.1	0.8	1.3	7.1	8.0	7.9
Cuba	—	1.7	3.5	—	0.5	1.2	6.0	6.4	7.0
West Germany	—	1.2	2.6	—	0.8	0.5	6.2	6.1	6.2
Hungary	0.2	1.4	2.1	—	0.1	0.5	4.8	5.1	5.5
Sweden	—	—	—	0.7	2.0	2.8	4.8	4.6	4.4
France	0.2	0.1	0.8	0.1	0.7	0.8	2.6	4.5	3.1
Japan	—	1.2	2.3	—	0.2	1.6	2.7	3.3	1.0
Yugoslavia	0.2	0.4	0.6	—	0.1	0.4	2.7	2.9	3.4
Belgium	—	—	—	—	0.2	0.1	1.3	2.0	2.5
United Arab Republic	0.2	0.7	0.7	0.2	0.6	0.1	1.6	1.6	1.4
The Netherlands	—	—	—	—	—	—	1.4	1.6	2.4
Austria	—	0.5	0.5	—	—	—	1.1	1.1	1.0
China	—	—	—	1.2	2.4	—	—	—	—
Total exported by USSR	2.9	17.8	43.4	5.1	15.4	21.1	66.8† 29.0‡	74.8† 30.3‡	76.2† 30.8‡

*In order of importance as of 1972.
†Crude oil.
‡Oil Products.

Source: Vneshnyaya torgovlya SSSR 1918-1966, Moscow 1967: 126-127; Vneshnyaya torgovlya SSSR za 1971 god, Moscow 1972: 68-69; Vneshnyaya torgovlya SSSR za 1972 god, Moscow 1973: 68-69.

TABLE 8.17

Main Countries Importing Soviet Hard
Coal and Anthracite

Country*	1955	1960	1965	1970	1971	1972
Bulgaria	—	—	2.5	5.1	6.0	5.7
East Germany	2.3	5.1	6.0	3.3	3.9	3.9
Czechoslovakia	—	1.1	2.9	2.7	2.9	2.9
Japan	0.1	0.5	1.2	2.9	2.5	2.5
Italy	0.2	0.5	1.0	2.0	1.8	1.6
France	0.6	0.8	1.6	1.5	1.4	1.2
Poland	—	0.8	1.2	1.1	1.2	1.2
Yugoslavia	0.3	0.9	1.1	1.1	1.1	1.1
Others	0.8	2.6	4.9	4.8	4.1	4.3
Total exported by USSR	4.2	12.3	22.4	24.5	24.9	24.2

*In order of importance as of 1971.

Sources: Vneshnyaya torgovlya SSSR 1918-1966, Moscow 1967: 126-27; Vneshnyaya torgovlya SSSR za 1971 god, Moscow 1972: 67-68; Vneshnyaya torgovlya SSSR za 1972 god, Moscow 1972: 68.

7,500 million kWh in 1972, and exports of gas rose from 4,600 million cu.m. to 5,100 million cu.m. in the same period. Exports of hard coal and anthracite (24.4 million tons) were slightly down, and exports of coke (4.5 million tons) much the same as in 1971.

In 1972 the USSR imported 9.7 million tons of hard coal and 0.6 million tons of coke, all from Poland. Total imports of crude oil were 7.8 million tons, mainly from Iraq (4 million), Libya (2 million), and Egypt (1 million). Only 1.3 million tons of oil products were bought abroad. Imports of natural gas from Iran and Afghanistan rose to 11,000 million cu.m. in 1972. This was double the amount of gas exported but sales of gas to Comecon nations and to Austria, Italy, West Germany, France, Finland, and Sweden were expected to grow enough to make the USSR a net exporter of natural gas as well as of oil and coal within the next few years.[16]

Of even greater potential importance were Soviet negotiations with Japan and the United States concerning possible participation in exploiting Siberian oil and gas. The Soviet view is that, since the United States is threatened with an energy crisis, while the USSR has the largest energy resources of any nation but lacks the capital and technology to fully exploit them, it would be to the advantage of both countries to cooperate in this sphere. The cost of extracting West Siberian gas, constructing pipelines to transport it to Murmansk, and then building liquefaction plants and special liquified-natural-gas tankers to ship the gas to the United States would be immense. American firms would be expected to supply their own equipment and technology on credit, with subsequent payment by the USSR through deliveries of natural gas. Early in 1974 it was by no means certain that the U.S. Export-Import Bank would be prepared to advance the loans necessary for cooperation on such a vast scale.[17]

Japan is also discussing cooperation in exploiting Soviet energy resources. In 1973 a proposal to help finance the development of the Tyumen oil fields and the extension of the existing oil pipeline from Irkutsk to Nakhodka was being considered. But Japanese hopes of importing up to 50 million tons a year from the USSR have been disappointed; Soviet sources have mentioned a maximum of only 25 million tons. Negotiations were also being conducted between the USSR and Japan on the subject of cooperation in the exploitation of natural gas in northern Sakhalin and Yakutia and of coking coal in southern Yakutia. Reserves still had to be confirmed by Japanese experts, and it still remained to agree on the quantity and cost of gas deliveries to the Japanese. It also had to be decided where the best site for liquefying the gas on the Pacific coast would be. De Kastri or Olga are possible bases, but delivery of gas is unlikely before 1980. In March 1974, Japan was pushing ahead with negotiations despite the ban on credit by the U.S. Export-Import Bank, which seemed likely to delay the involvement of potential U.S. partners.[18]

Although it is not the purpose of this work to speculate on future Soviet export potential, certain influential factors may be mentioned here.

On the present showing, output of oil and gas in West Siberia can be expected to continue to increase rapidly in coming years; however, if a determined effort is made by the government to raise the living standard of the population to that of developed nations in the West, exports of energy will have to be restricted to the minimum that can cover the cost of importing Western technology.

It is possible that the USSR could increase its oil exports to Europe by importing 50 to 100 million tons from the North Rumaila fields in Iraq, which Soviet experts have been helping to develop. As has been shown, this policy has already been followed in the importing of natural gas from Afghanistan and Iran.

An important factor that influences Soviet exports to the West is the question of how much energy the USSR will be prepared to sell to the countries of Eastern Europe, which at present suffer from an energy deficit. Table 8.18 shows that only the USSR is a net exporter

TABLE 8.18

Comecon Trade in Fuels, 1970
(Million tons)

	Coal	Coke	Crude Oil	Oil Products
Bulgaria	+4.9	+0.5	- 0.1 + 5.7	- 0.2 +2.6
Czechoslovakia	- 3.0 +4.5	-2.5	- 0.1 + 9.8	- 0.7 +0.6
East Germany	+7.9	+3.1	+10.3	- 1.2
Hungary	+2.0	+1.3	- 0.3 + 4.4	- 0.9 +0.9
Poland	-28.8 +1.1	-2.3	+ 7.0	- 1.3 +2.4
Romania	+0.7	+2.6	+ 2.3	- 5.1
USSR	-24.4 +7.0	-4.2	-66.8 + 3.5	-29.0 +1.1*

- exported
+imported
*crude and products

Source: Statisticheskii ezhegodnik stran-chlenov SEV 1971, Moscow 1972: 348-376.

TABLE 8.19

Production of Fuels and Electricity in
Comecon Countries, 1970

	Electricity* (Th. Mill. Kwh)		Hard Coal	Brown Coal	Coke	Oil	Natural Gas (Th. Mill. Cu. M.)
				(Million Tons)			
Bulgaria	20	(2)	—	29	1	—	—
Czecho-slovakia	45	(4)	28	81	12	—	1
East Germany	68	(1)	1	261	8	—	—
Hungary	15	—	4	24	1	2	3
Mongolia	1	—	—	2	—	—	—
Poland	65	(2)	140	33	17	—	5
Romania	35	(3)	6	14	1	13	24
USSR	740	(124)	433	145	75	349	184

*Of which hydro–electricity shown in parentheses.
— Less than 0.5.

Source: Statisticheskii ezhegodnik stran-chlenov SEV 1971, Moscow 1972: 77-79.

of both coal and oil. Poland exports large quantities of coal but has
to import oil. The other countries, including Cuba, which joined
Comecon in 1972, are heavily dependent on the USSR for their energy
supplies. A comparison of Table 8.18 with Table 8.19 shows that only
Romania produces enough oil and gas to meet its own requirements.
Poland, East Germany, and Czechoslovakia produce large amounts of
coal but will have to continue to import a growing quantity of oil. If
the Soviet government intends to pay for Western technology by exporting oil and gas to the West rather than to Comecon countries, the latter
will probably have to seek a growing proportion of their supplies in
the Middle East.

NOTES

1. Sovetskii Soyuz—obshchii obzor (The Soviet Union—general
survey), Moscow 1972: 488.
2. Melnikov, N. V., Mineralnoe toplivo (Mineral fuel), Moscow
1971: 173.
3. Ibid.: 182.
4. Energeticheskie resursy SSSR: toplivno-energeticheskie
resursy (Energy resources of the USSR: fuel resources), Moscow
1968: 508-10.
5. Ibid.: 511-20; Evdokimenko, A. I., Kosterin, V. V., Prirodnyi
gaz v tsvetnoi metallurgii (Natural gas in the nonferrous metal industry), Moscow 1972: 14-20; Maslakov, D. I., "Gaz v narodnom
khozyaistve soyuznykh respublik" (Gas in the economy of the union
republics), Gazovaya promyshlennost no. 12, December 1972: 6-7.
6. Komsomolskaya pravda, 31 March 1973: 4; Trud, 25 January
1973: 4; Izvestiya, 25 January 1973: 6.
7. Melnikov, op. cit.: 195-203; Ravich, M. B., Toplivo i effektivnost ego ispolzovaniya (Fuel and its efficient utilization), Moscow
1971: 202; SSSR v tsifrakh v 1971 godu (The USSR in statistics for
1971), Moscow 1972: 92-93; 151; Tolkachev, A. S., et al., Osnovnye
napravleniya nauchno-tekhnicheskogo progressa (The basic directions
of scientific and technological progress), Moscow 1971: 111-25.
8 Rylskii, V. A., Ekonomika mezhraionykh electroenergeti-
cheskikh svyazei v SSSR (The economics of interregional energy links
in the USSR), Moscow 1972: 9-14.
9. Planovoe khozyaistvo no. 4, April 1973; Planovoe khozyaistvo no. 5, May 1973; Soviet Analyst 2, no 16, 2 August 1973: 4.
10. Evdokimenko, op. cit.: 13-14.
11. Rylskii, op. cit.: 15-16.
12. Narodnoe khozyaistvo SSSR 1922-1972 (National Economy
of USSR 1922-1972), Moscow 1972.

13. Vneshnyaya torgovlya SSSR za 1972 god (Foreign trade of the USSR in 1972), Moscow 1973: 68.

14. Ibid.: 69.

15. Ibid.: 69, 100; Evdokimenko, op. cit.: 14-15.

16. Ekonomicheskaya gazeta no. 17, April 1973: 20-21; Vneshnyaya torgovlya SSSR za 1972 god, Moscow 1973: 41.

17. Shershnev, E. S., "Sovetsko-amerikanskie ekonomicheskie otnosheniya, ikh perspektivy" (Soviet-American economic relations: prospects), S. Sh. A. (USA) no. 1, January 1973: 25-26.

18. Pishchik, B., "Japan in the Changing World," New Times no. 23, June 1973: 12-13; Saeki, K., "Japan and Siberia," Problems of Communism, May-June 1972: 5-11; BBC monitoring service, USSR Weekly Economic Report, 15 June 1973; Ibid.: 21 August 1973; Problemy Dalnego Vostoka, no. 3, 1973: 70; International Herald Tribune, 21 March 1974: 7.

The USSR has reserves of all the major sources of energy that are sufficient to satisfy its own requirements and to ensure that output can continue to increase over the foreseeable future. The basic problem facing Soviet energy specialists and economic planners is that some 90 percent of the total energy resources lie east of the Urals, while population and industry are concentrated in the Urals and in the European parts of the USSR.

The USSR has a greater share of the world's energy resources than any other country: in 1972 it could claim 57 percent of the coal, 45 percent of the natural gas, 60 percent of the peat, 46 percent of the oil shale, 12 percent of the hydroelectric power, and 37 percent of the oil-bearing area of the world.

The total output of fuel in the USSR has increased from 311 million conventional fuel tons in 1950 to 1,354 million tons in 1972. In the same period the production of coal more than doubled, but its proportion of the fuel balance has dropped from 66 percent to 34 percent; output of oil in 1971 was over ten times the output in 1950, and its proportion of the fuel balance had increased from 17 percent to 42 percent; natural gas production in 1972 was some 35 times the 1950 level, and its proportion of the fuel balance had increased from 2 percent to 20 percent.

This trend is expected to continue. By the year 2000 natural gas could account for as much as 35 percent of the Soviet energy balance, oil about 26 percent, and coal under 20 percent. Energy requirements will therefore still be mainly satisfied by fossil fuels, although nuclear energy is expected to increase its share from less than 1 percent in 1970 to about 16 percent by the end of the century.

The USSR is second only to the United States in total energy output. At the beginning of the 1970s the Soviet Union was producing about three-quarters as much oil and one-third as much natural gas as was produced in the United States, but since the ratio of reserves to production seems considerably better in the USSR than in the United States, this situation is likely to change in favor of the Soviet Union in the decades to come. In actual tonnage of coal extracted, the USSR had already overtaken the United States, but since a larger proportion of Soviet output consisted of brown coal, the United States was still ahead in the calorific value of the coal produced. Through greater reliance on nuclear energy and by importing oil, the United States seems likely to retain its considerable lead over the USSR in electric power production.

At present the major oil-producing areas in the USSR are the Volga-Urals, Tyumen, and North Caucasus in the RSFSR; Mangyshlak in Kazakhstan; Turkmenia; and the Ukraine. Natural gas is extracted mainly in the Ukraine, Uzbekistan, North Caucasus (RSFSR), and Turkmenia. The major coal producers are the Donets, Kuznetsk, Karaganda, Pechora, Moscow, Ekibastuz, and Kansko-Achinsk basins.

Production of peat and oil shale has also been rising, and although taken together they supply only 2 percent of the total Soviet fuel requirement, they form a much more significant part of the energy balance in certain specific areas: peat, for instance, in Belorussia; and oil shale in Estonia.

The distribution of the energy-producing industries is in sharp contrast with the distribution of resources. The European part of the USSR, with only 10 percent of the USSR energy resources, produces about 70 percent and consumes 80 percent of the total output. Even if the utilization of local sources of energy is increased to the maximum that is economically viable, it will still be necessary to import fuel and electricity from east of the Urals.

The information that is available about the quantity, location, and present utilization of Soviet energy sources would therefore suggest that certain lines of development will be followed in coming years.

Total energy consumption will continue to expand rapidly, with the more economical fuels, natural gas and oil, increasing their share in the fuel mix. The share of coal will fall even further although actual output will increase, mainly from cheaper opencast mines.

Production will expand, chiefly in the eastern regions; oil in Tyumen oblast (West Siberia) and Mangyshlak (Kazakhstan); gas in Tyumen and Yakutia (East Siberia) and in the Central Asian republics; coal in Siberia and Kazakhstan. Large energy bases are being created in these areas that will not only satisfy local requirements but will produce a surplus for transmission to the Urals and further west. Large quantities of oil and gas will probably be exported to pay for foreign technology.

This long-distance transportation will involve further construction of large-diameter oil and gas pipelines and the development of direct current high-voltage transmission lines.

The distribution of energy demands will be improved by expanding energy-intensive industries in appropriate eastern regions.

In the European part of the USSR the older oil and gas fields will be developed more intensively and at greater depths. There will be increased exploration of offshore fields in the Caspian, Black, and Baltic seas and of new fields in the Ukraine and Belorussia. The Donets and Moscow coal basins will be exploited more intensively, with even comparatively poor seams being worked because of high local demands.

Production of oil shales will expand to help satisfy the energy requirements of Estonia and the Leningrad area. Further construction of electric power stations based on new large-scale peat enterprises is advisable in western and central parts of the USSR. The consumption of peat in agriculture and the chemical industry will continue to rise.

Since only a fraction of the potential power of Soviet rivers has been tapped, many more hydroelectric stations can be built. Tidal and geothermal energy are other possible areas for expansion, but their contribution to the total Soviet energy balance in this century will remain insignificant.

Nuclear power stations will play an increasingly important part in satisfying the energy requirements not only of remote areas in the Far North but also in the energy-hungry regions in the industrial European parts of the USSR.

Soviet energy specialists, however, are not concentrating solely on increasing output. Considerable attention is being paid to problems concerning energy conservation and the more rational utilization of power resources. The present low efficiency rates of fuel utilization in the economy (about 33 percent) and the considerable losses during the extracting, processing, and transporting of fuels reduce actual useful consumption rates to around 15 percent. New complex systems of combining power production with the technological processing of fuels are being developed to help improve fuel utilization.

While the Soviet Union has no need to fear an absolute energy shortage, improving the efficiency of fuel utilization and rationalizing the distribution of energy production and consumption will continue to pose complex problems for Soviet planners in the decades to come.

DEFINITIONS

1. Soviet Ranks of Coal

Symbol	Name and Description	Percentage of Volatiles	Percentage of Carbon
A	Anthracite (antratsit). Hard; jet-black color; clean to handle; non-coking; burns with short, pale-blue flame.	4	96
PA	Semi-anthracite (poluantratsit). Similar to anthracite but more friable; burns with yellow flame.	8	94
T	Lean coal (toshchii). Dry, steam coal; semibituminous; dark grey color; gives crumbly coke.	13	92
OS	Lean caking (otoshchenyi-spekayush-chiisya).	22	90
PS	Steam caking (parovichno-spekayush-chiisya). Dry steam; black color with grey shades; fairly brittle.		
K	Coking (koksovyi). Deep black color; very brittle; gives hard, strong coke.	25	88
Zh	Fat (zhirnyi). Metabituminous; medium volatile.	32	87
PZH	Steam fat (parovichno-zhirnyi). Deep black color with slight brown shades; extremely brittle.		

Symbol	Name and Description	Percentage of Volatiles	Percentage of Carbon
G	Gas (gazovyi). Black color with brown shades.	40	83
D	Longflame (dlinnoplamennyi). Black or dark brown color.	42	80
DB	Brown longflame (dlinnoplamennyi-buryi) between brown and hard coal; subbituminous.		
B	Brown (buryi) lignite, hydrogenous layered structure; fragments of wood.	50	70

2. General Glossary

$A + B + C_1 + C_2$ Categories for dividing mineral resources in accordance with the extent to which they have been explored. For oil and gas (and categories $D_1 + D_2$), see pages 17-18, 90. For coal, see page 157.

ASSR   Autonomous Soviet Socialist Republic. Administrative area giving some political recognition to a minority nationality. There are twenty of them, all subordinate to the Union Republic in which they are situated.

Atmosphere   A pressure of 1 kg. per sq. c. or of 14.7 lb. per sq. in. (in U.S. measure).

Balance resources   A term for describing those natural resources that can be utilized economically with regard to industrial requirements such as the quality of the raw material and the conditions of extraction. The exploitation of extra-balance resources is not at present economically expedient because they lie in thin beds, are difficult to mine, or require complex processing.

CC	Central Committee.
Comecon	Council of Economic Mutual Assistance. Includes USSR, Bulgaria, Czechoslovakia, East Germany, Hungary, Mongolia, Poland, Rumania, and (since 1972) Cuba.
Condensate	See "Gas."

Conventional fuel ton (CFT)

A useful term for comparing different forms of energy by converting them into their equivalent calorific values. One CFT has a calorific value of 7,000 k.cal./kg. Fuels are normally converted to CFT in Soviet sources by using the following standard calorific values:

	k.cal./kg.
hard coal	7,000
brown coal	2,500
oil	10,000
natural gas	8,800
peat	2,800
oil shale	1,900

An alternative method of conversion is also used:

	CFT
one ton of brown coal	.04
one ton of hard coal	1.0
one ton of oil	1.4
1,000 cu.m. of gas	1.3

CPSU

Communist Party of the Soviet Union

Gas

Gas from natural gas fields is composed mainly of methane (up to 99 percent) and contains few heavier hydrocarbons. Its minimum calorific value is 7,000 to 8,500 k.cal./ cu.m. Gas from condensate fields has more heavy hydrocarbons and contains condensate, which is extracted. Its minimum calorific value is 8,000 to 10,000 k.cal./cu.m. Incidental gas from oil fields contains much less methane (41 percent in Bashkir gas, 65.5 percent in Krasnodar gas, 91.8 percent in Azerbaidzhan gas) but always has a high

content of heavy hydrocarbons. It is best
used in the production of petrol and chemicals.

Geological reserves	The total of all categories of explored reserves plus predicted reserves.
GOELRO	State Commission for the Electrification of Russia
GOSPLAN	State Planning Commission
Incidental gas	Gas produced during the extraction of oil. See "Gas."
k.cal./kg.	kilocalorie per kilogram.
Kolkhoz	Collective farm.
Kopek	One hundredth of a ruble.
Krai	Administrative subdivision, subordinate to the Union Republic in which it is situated. Some contain lesser political subdivisions based on nationality groups (autonomous oblast or national okrug).
Oblast	An administrative subdivision. Many oblasts are larger in area than the smallest of the Union Republics, but they are always subordinate to the Union Republic in which they are situated.
Raion	Lowest level of administrative area.
RSFSR	Russian Soviet Federative Socialist Republic. The largest of the fifteen Union Republics.
Ruble	Unit of Soviet currency. At the February 1974 official rate of exchange, £1 equalled 1.72 rubles and $1 equalled 0.79 rubles. The average industrial wage in the USSR was approximately 135 rubles a month.
Sovkhoz	State farm

SSR   Soviet Socialist Republic, of which there are
                        fifteen in the USSR.

USSR   Union of Soviet Socialist Republics (Soviet
                        Union). Consists of fifteen union republics.

COMPARATIVE COSTS OF EXTRACTING FUELS, BY AREA

1. GAS (in rubles per thousand cu.m.)

East Ukraine	6.6
Stavropol krai	5.3
Krasnodar krai	5.8
Azerbaidzhan	15.7
Volgograd oblast	11.5
Tyumen oblast	2.6
Turkmenia	4.9
Uzbekistan	6.6
Tomsk oblast	7.7
Yakut ASSR	6.2

2. OIL (in rubles per ton)

Komi ASSR (mine)	15.4
Perm and Orenburg oblast	14.8 to 15.0
Bashkir ASSR	12.7
Tatar ASSR	8.7
Kuibyshev oblast	11.4
Stavropol and Krasnodar krai	16.3 to 21.8
Chechen-Ingush ASSR	20.0
Tyumen oblast	8.8
Irkutsk oblast	13.0
Ukraine	13.3
Azerbaidzhan	26.8
Turkmenia	22.8
Kazakhstan (Mangyshlak)	9.5

3. COAL (in rubles per ton)

Donets basin
Old mines	10.5
New mines	12.1

Kuznetsk basin
Old mines   6.3
Reconstructed opencasts   4.9
New opencasts   5.5
Pechora basin
Old mines   8.9
New mines   11.3
Moscow basin
Old mines   5.9
New mines   7.8
Old opencasts   2.9
Kansko-Achinsk basin
New opencasts   0.9
Ekibastuz basin
Old opencasts   1.6
New opencasts   2.0
Far East fields
Old mines   10.4
New mines   9.5
New opencasts   4.2

4. PEAT (in rubles per ton)

Northwest region   3.8
Ural region   3.5
West Siberian region   8.8
Far East region   4.0
Southwest region   4.0
Baltic region   4.4
Belorussia   4.4

5. OIL SHALE (in rubles per ton)

Baltic Basin
Mine   4.0
Opencast   3.1
Volga Region
Mine   4.8
Opencast   3.0

Source: Melnikov, N. V., Mineralnoe toplivo, Moscow 1971:
178-80.

270

COMPARATIVE COSTS OF TRANSPORTING ENERGY,
BY DISTANCE AND FORM OF TRANSPORT

Form of Transport	Yearly Throughput		Cost by Distance (in rubles per conventional fuel ton)			
			1,000 km.	2,000 km.	3,000 km.	4,000 km.
	(in thousands of millions of cu.m.)	(in millions of conventional fuel tons)				
Gas pipeline (in mm. diameter)						
1,420	28	33	2.3	4.9	7.8	10.8
2,020	56	66	1.9	4.0	6.5	8.9
2,520	85	100	1.6	3.3	5.2	7.1
	(in millions of tons)	(in millions of conventional fuel tons)				
Oil pipeline (in mm. diameter)						
780	18	25	0.6	1.3	1.9	2.6
820	27	38	0.5	1.0	1.5	2.1
1,020	48	67	0.4	0.8	1.2	1.6
	(in millions of megawatt hours)	(in millions of conventional fuel tons)				
Direct current very high tension transmission line (in KV)						
$\pm$ 750	42	15	—	5.6	7.1	8.4
$\pm$ 1100	90	30	—	—	5.2	6.2
	(in millions of tons)	(in millions of conventional fuel tons)				
Coal By Railway (in k.cal./ kg.)						
6,000	100	86	1.9	4.0	5.6	7.0
3,500	100	50	3.2	6.8	9.6	12.0

Source: Melnikov, N. V., Mineralnoe toplivo, Moscow 1971: 182.

COMPARATIVE ENERGY COSTS IN AREA OF CONSUMPTION
(IN RUBLES FOR EVERY TEN CONVENTIONAL FUEL TONS)

Source of Energy	Extraction Area	Leningrad (Northwest)	Moscow (Center)	Riga (Baltic)	Minsk (Belorussia)	Voronezh (Central Black Earth)	Gorkii (Volgo-Vyatka)	Donets-Dnepr	Sverdlovsk (North Ural)	Magnitogorsk (South Ural)
Donbass coal	143	174	166	178	170	155	185	151	—	—
Pechora coal	134	176	174	—	—	185	—	—	—	—
Moscow coal	209	—	231	—	—	—	—	—	—	—
Kuzbass coal										
Mine	110	181	172	190	180	177	166	—	146	146
Opencast	77	163	152	174	163	159	144	—	122	125
Ekibastuz Coal	28	—	—	—	—	117	—	—	67	58
Kansko-Achinsk Coal										
Dried fines	42	141	125	148	138	130	125	145	100	103
Semicoke fines	43	117	105	122	115	107	105	120	87	89
Semicoke nuts	45	119	110	127	113	116	105	125	88	90
Milled Peat	—	129	125	129	125	152	108	—	112	—
Oil Shale										
Mine	—	—	—	110	—	—	—	—	—	—
Opencast	—	—	—	84	—	—	—	—	—	—
Fuel Oil	—	89	86	92	98	116	86	91	106	119
Tyumen Gas	22	78	77	100	88	—	86	—	66	—
Uzbekistan Gas	57	—	99	—	—	82	—	—	—	82
Stavropol Gas	46	—	—	—	—	—	—	62	—	—
Turkmenistan Gas	42	—	85	—	—	96	—	91	—	—
Electric Power										
by DC transmission										
1500 kV (± 750 kV)	—	—	—	—	—	59[a]	89[b]	103[b]	—	—
2200 kV (± 1100 kV)	—	—	—	—	—	—	—	80[b]	—	—

[a] from Ekibastuz
[b] from Itat

Source: 1968 Data from State Committee for Science and Technology of the Academy of Sciences, USSR; Melnikov, N.V., Mineralnoe toplivo, Moscow 1971: 183.

A NOTE ON WOOD AS FUEL

In 1860 wood formed about 60 percent of the world's fuel balance. The industrial revolution brought a sharp fall in its importance as a fuel. In 1900 world requirements for wood amounted to 167 million tons of conventional fuel, making its share in the fuel structure 17.6 percent. In 1970 the world demand for wood reached 250 million conventional fuel tons, but its share in the fuel structure dropped to only 3.4 percent.

The USSR has one-fifth of the world's forests, stretching over 7.5 million sq.km., or a third of Soviet territory. Total timber resources are estimated at 80,000 million cu.m. and are growing by almost 800 million cu.m. a year.

Consumption of wood as a fuel in the USSR rose from 9.7 million conventional fuel tons in 1913 to 33.5 million tons in 1965, but its share in the Soviet fuel mix dropped from 20.1 percent to 3.5 percent in the same period. By 1971 consumption had dropped to 26.6 million tons and the share of wood in the fuel mix to 2.1 percent. Since these figures refer to wood consumed by official agencies and do not include wood burned by private individuals, total real consumption was much higher.

Wood is still very important in the fuel balance of certain forest areas. In 1965 it contributed 34 percent of the fuel consumed in the Karelia ASSR, 28 percent in Arkhangelsk oblast, 26 percent in Novgorod oblast, 15 percent in Pskov oblast, 10 percent in Komi ASSR, 8 percent in Leningrad oblast, 8 percent in Vologda oblast, and 6 percent in Murmansk oblast. The share of wood in the fuel balance of the Northwest economic region was 9 percent and in the fuel balance of the Ural economic region was 5 percent.

Burning wood is a wasteful use of a valuable raw material, and the amount of wood consumed as a fuel is expected to continue to drop.

Sources: Ravich, M. B., Toplivo i effektivnost ego ispolzovaniya (Fuel and its efficient utilization), Moscow 1971: 117-18; Ezhegodnik "SSSR-72" (USSR in 1972), Moscow 1972: 24-25; Narodnoe khozyaistvo SSSR 1922-1972 (National economy of the USSR 1922-1972), Moscow 1972: 162.

ENGLISH

Campbell, R. W. The Economics of Soviet Oil and Gas, Baltimore
    1968.

Ebel, R. E. Communist Trade in Oil and Gas, New York 1970.

Hodgkins, J. A. Soviet Power: Energy Resources, Production and
    Potential, London 1961.

RUSSIAN

General Background Books

Baranov, A. N., et al. Atlas SSSR, Moscow 1969.

Bolshaya Sovetskaya Entsiklopediya, 2nd Ed., Moscow 1949, 1958;
    3rd Ed., Moscow 1970.

Drobizhev, V. Z., et al. Istoricheskaya geografiya SSSR, Moscow
    1973.

Kalesnik, S. V., et al. Sovetskii Soyz — Geograficheskoe opisanie
    v 22-kh tomakh, Moscow 1967, 1972.

Khrushchev, A. T. Geografiya promyshlennosti SSSR, Moscow
    1969.

Nikitin, N. P., et al. Ekonomicheskaya geografiya SSSR, Moscow
    1966.

Ponomarev, B. N., et al. Istoriya SSSR v 12-i tomakh, Moscow
    1967.

Stroev, K. F., et al. Ekonomicheskaya geografiya SSSR, Moscow
    1972.

Vosnesenski A. N., et al. Atlas razvitiya khozyaistva i kultury
    SSSR, Moscow 1967.

Specialist Books on Energy Resources and their Utilization

Andreev, V.. Atom sluzhit miru, Moscow 1968.

Bakirov, A. A., Ryabukhin, G. E. Neftegazonosnye provintsii i oblasti SSSR, Moscow 1969.

Bratchenko, B. F., et al. Ugolnaya promyshlennost SSSR 1917-1967, Moscow 1969.

Driatskaya, Z. V., et al. Nefti SSSR, Moscow 1971.

Kortunov A. K., et al. Gazovaya promyshlennost SSSR, Moscow 1967.

Kuznetsov, K. K., et al. Ugolnye mestorozhdeniya dlya razrabotki otkrytym sposobom, Moscow 1971.

Lvov, M. S. Resursy prirodnogo gaza SSSR, Moscow 1969.

Matveev, A. M., et al. Torf v narodnom khozyaistve, Moscow 1968.

Melnikov, N. V., et al. Energeticheskie resursy SSSR—Toplivno-energeticheskie resursy, Moscow 1968.

Melnikov, N. V. Mineralnoe toplivo, Moscow 1971.

Melnikov, N. V. Toplivno-energeticheskie resursy SSSR, Moscow 1971.

Neporozhnii, P. S. Elektrifikatsiya SSSR 1917-1967, Moscow 1967.

Ravich, M. B. Toplivo i efektivnost ego ispolzovaniya, Moscow 1971.

Rostovtsev, M. I., Runova, T. G. Dobyvayushchaya promyshlennost SSSR, Moscow 1972.

Rylskii, V. A. Ekonomika mezhraionnykh elektroenergeticheskikh svyazei v SSSR, Moscow 1972.

Sidorenko, A. V., et al. 50 let Sovetskoi geologii, Moscow 1968.

Tolkachev, A. S., et al. Osnovnye napravleniya nauchno-tekhni-cheskogo progressa, Moscow 1971.

Vilenskii, M. A. <u>Po Leninskomu puti sploshnoi elektrifikatskii</u>, Moscow 1969.

Volkov, T. M., et al. <u>Razrabotka i ispolzovanie zapasov goryuchikh slantsev</u>, Tallin 1970.

Voznesenskii, A. N., et al. <u>Energeticheskie resursy SSSR — Gidroenergeticheskie resursy</u>, Moscow 1967.

FUEL AND ENERGY JOURNALS

Atomnaya energiya

Gazovaya promyshlennost

Gidrotckhnicheskoe stroitelstvo

Neftyanik

Neftyanoe khozyaistvo

Stroitelstvo truboprovodov

Torfyanaya promyshlennost

Ugol

STATISTICAL YEARBOOKS

<u>Narodnoe khozyaistvo SSSR</u>, Moscow.

<u>Statistical Abstract of United States</u>, Washington.

<u>Statisticheskii ezhegodnik stran-chlenov Soveta Ekonomicheskoi Vzaimopomoshchi</u>, Moscow

<u>SSSR i soyuznie respubliki</u>, Moscow.

<u>SSSR v tsifrakh</u>, Moscow.

<u>United Nations Yearbook of International Trade Statistics</u>, New York.

<u>United Nations Statistical Yearbook</u>, New York.

Vneshnyaya torgovlya SSSR, Moscow.

NEWSPAPERS

Ekonomicheskaya gazeta

Izvestiya

Komsomolskaya Pravda

Pravda

Sotsialisticheskaya industriya

Sovetskaya Rossiya

Trud

SWB—Summary of World Broadcasts. The USSR Weekly Economic
    Report: Fuel and Power. Published by the monitoring service
    of the British Broadcasting Corporation, Caversham Park,
    Reading.

IAIN F. ELLIOT is Lecturer in Russian Studies at Brighton Polytechnic, England. He is editor of the fortnightly newsletter <u>Soviet Analyst</u>, and is a member of the National Association for Soviet and East European Studies. Since 1969 he has been working with Dr. V. S. Balashov of Bradford University on a research project examining the development of power resources in the USSR.

From 1967 to 1968 Mr. Elliot was a research scholar at the University of Leningrad. He has travelled extensively in the Soviet Union.

Mr. Elliot is a graduate of Glasgow University, where he was awarded two degrees, the first in the Russian, French, and German languages, and the second in Soviet and East European studies.

RELATED TITLES
Published by
Praeger Special Studies

ISRAEL AND IRAN: Bilateral Relationships and
Effect on the Indian Ocean Basin
Robert B. Reppa, Sr.

MIDDLE EAST OIL AND U.S. FOREIGN POLICY
Shoshana Klebanoff

THE PRICING OF CRUDE OIL: Economic and
Strategic Guidelines for an International Energy
Policy
Takai Rifai

REGULATING THE NATURAL GAS INDUSTRY:
Pipelines, Conglomerates, and Producers
Samuel Blitman